MadCap Flare for Programmers

A guide to getting the most from Flare

Thomas Tregner

David Owens

MadCap Flare for Programmers

A guide to getting the most from Flare

Copyright © 2015 Thomas Tregner and David Owens

All rights reserved. No part of this book may be reproduced or transmitted in any form or by any means without the prior written permission of the copyright holder, except for the inclusion of brief quotations in a review.

Credits

Cover image: NASA Solar Dynamics Observatory, solar flare, public domain.

Disclaimer

The information in this book is provided on an "as is" basis, without warranty. While every effort has been taken by the authors and XML Press in the preparation of this book, the authors and XML Press shall have neither liability nor responsibility to any person or entity with respect to any loss or damages arising from the information contained herein.

This book contains links to third-party web sites that are not under the control of the authors or XML Press. The authors and XML Press are not responsible for the content of any linked site. Inclusion of a link in this book does not imply that the authors or XML Press endorse or accept any responsibility for the content of that third-party site.

Trademarks

XML Press and the XML Press logo are trademarks of XML Press.

All terms mentioned in this book that are known to be trademarks or service marks have been capitalized as appropriate. Use of a term in this book should not be regarded as affecting the validity of any trademark or service mark.

XML Press
Laguna Hills, California
http://xmlpress.net

First Edition
ISBN: 978-1-937434-25-0 (print)
ISBN: 978-1-937434-26-7 (ebook)

Table of Contents

Foreword vii
1. Introduction 1
    Audience for this book 3
    Purpose of this book 4
    Structure of this book 4
2. Flare Projects 7
    Project files 7
    Root folder and explorers 8
3. Flare, Topic-based Authoring, and DITA 13
    Topic-based authoring 13
    DITA 13
    Topic-based authoring in Flare 14
    Flare and other tools and standards 15
4. Flare Topics 19
    Flare topics overview 19
    Topic structure 24
    Topic output 30
    Import behavior 31
    Manipulating topic files 32
    Topic code samples 33
5. Flare Tables of Contents 43
    TOC structure 43
    Manipulating TOCs 47
    TOCs in output 48
    TOC code samples: creating and modifying a TOC 53
6. Flare Indexes, Glossaries, and Search 65
    Flare indexes 65
    Flare glossaries 70
    Search 72
7. JavaScript in Flare Topics, Master Pages, Snippets, and Skins 79
    Toolbar JavaScript through the Skin Editor 79
    Scripts in topics, master pages, and snippets 81
8. Document Automation and Batch Files 83
    Batch files 83
    Creating a manageable template for batch files 84

Automating imports 85
Building subsystems in parallel 86
Batch file generator utility 86
9. Connecting Applications to Flare Help Outputs 91
HTML5 tri-pane help 91
WebHelp 97
CHM and HTML Help 97
DotNet Help 97
Eclipse Help 98
10. Flare Application and Flare Plugins 99
Creating a plugin project 100
Building and deploying a plugin 109
Menus 111
Editors 115
Documents 118
Insert element example 121
11. Formatting and Pasting Code Samples into Flare 125
Plugin for pasting code samples 127
12. Strategies for Combining Generated and Authored Content 131
Strategy: Don't touch generated content 131
Strategy: Generate once and edit freely 133
Strategy: Generate content but only keep the topics 134
Strategy: Generate content as snippets 134
Strategy: Generate topics and snippets 135
Conclusion 135
Acknowledgments 137
A. Batch File Generator Utility 139
B. Element List 149
C. Glossary 155
About the Authors 163
Index 165

List of Examples

4.1. CSS to set `mc-master-page` for the `<html>` element 23
4.2. XML for a simple Flare topic 24
4.3. Code to insert a snippet in a topic 27
4.4. Topic with a condition applied to an element 27
4.5. Examples of proxy markup in a source topic 29
4.6. Generated proxy markup in an HTML5 topic 29
4.7. Flare Project Import File markup 32
4.8. XSLT that transforms every `<h1>` element into an `<h2>` element 35
4.9. Batch file to run `msxsl.exe` 36
4.10. Transformed Flare topic markup (output from Example 4.8) 36
4.11. XSLT that finds a particular attribute-value pair and replaces the value 37
4.12. C# console application to change a topic title 38
4.13. Visual Basic .NET console application that appends a table to a topic 39
4.14. JavaScript fragment to underline content with `class="ProductsFlare"` 41
5.1. XML for a TOC with a single link 43
5.2. XML for a TOC with a variety of node types 44
5.3. XML for a TOC with a sub-TOC 44
5.4. Sub-TOC or a standalone TOC (`ChildTOC.fltoc`) 45
5.5. TocEntry with a bookmark indicating an HTML5 target 46
5.6. Script tags in an HTML5 main page (as of version 9) 49
5.7. `Toc.js` file 50
5.8. HTML5 script tags for `require.min.js` and `require.config.js` 51
5.9. Code to traverse the tree in `Toc.js` 52
5.10. XSLT that appends "_topic" to each entry in a TOC 53
5.11. Build TOC from CSV list 54
5.12. Visual Basic logic for custom TOC sorter 56
5.13. Top level HTML5 output TOC sort with JavaScript 60
5.14. An approach to creating Flare topics and TOCs with Java 60
6.1. XML for a `<CatapultIndexLinkSet>` element with no entries 66
6.2. WordDetail class 67
6.3. Module to create an Auto Index Set 67

6.4. XML markup for a glossary file 70
6.5. .NET code to add a term to the glossary 71
6.6. JavaScript to sort an HTML5 glossary in reverse alphabetical order 72
6.7. XML markup for a search data file (`Search.xml`) 73
6.8. XML markup for a search chunk data file (`Search_Chunk1.xml`) 74
6.9. XML markup for a search filter 75
6.10. Simple input field to open a qualified URL for HTML5 search 76
6.11. C# tool to create multiple targets to test search settings 77
7.1. JavaScript to add keyup behavior for the search field 81
8.1. Batch file to run `madbuild.exe` 84
8.2. Batch file to check out, process, and check in files 85
8.3. Parallel build using the `start` command 86
9.1. JavaScript button that calls `MadCap.OpenHelp` 93
9.2. Sample alias file (`AliasFile.flali`) 94
9.3. Sample header file (`HeaderFile.h`) 94
9.4. Java methods for opening URLs 95
9.5. Sample properties file (`HeaderFile.properties`) 96
10.1. C# class file generated by Visual Studio (slightly adjusted) 101
10.2. C# `Class1.cs` file with assembly references added 104
10.3. Stubbed out C# implementation of IPlugin interface (spacing adjusted) 106
10.4. C# Code for creating a toolbar, ribbon, and button command (version 10) 113
10.5. Modify the `Execute()` function to create the toolbar and ribbon 114
10.6. C# Code for creating a toolbar, ribbon, and button command (version 9) 114
10.7. API details code snippet for gathering information returned by the plugin API 117
10.8. Open file dialog with a plugin 119
10.9. Insert element example 121
11.1. Sample C# plugin for pasting code samples 127
A.1. C# code for the Flare batch commands manager 139

Foreword

MadCap Software is dedicated to providing leading-edge software that addresses the many ways people obtain their information, whether it is in print, on the Web, or on a mobile device. Technical communicators, marketing managers, clinical documentation specialists, user assistance managers, and policy specialists across a wide range of industry verticals use our software to create, manage, and publish content quickly and efficiently.

Flare for Programmers is a valuable resource for anyone looking to adapt and expand on the robust capabilities of MadCap Flare. Since Flare's source files are all based on open standards such as XML, JavaScript, and CSS, the examples and techniques covered in this book are not only useful in the MadCap Flare environment, but can also be applied to web development best practices, file management, and XML manipulation.

With step-by-step examples for processes including dynamic content generation and plugin API development, this book offers the most thorough overview of advanced Flare development available to date.

Anthony Olivier,
CEO of MadCap Software

CHAPTER 1
Introduction

MadCap Flare[1] is a Microsoft Windows-based tool for maintaining content as a single source and publishing that content to multiple channels. The MadCap Flare user interface enables you to manipulate your content and produce a wide variety of deliverables, including help systems, PDF, and HTML. Flare lets you configure the user interface using windows, panes, and tabs. You can choose between a classic menu interface – the Toolstrip Menu – and a ribbon interface. Flare's window layout is customizable, allowing you to create a tailored work space with multiple editors and explorers (see Figure 1.1).

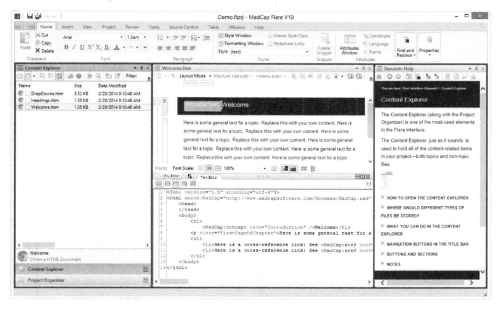

Figure 1.1 – Flare user interface

One of the main strengths of Flare is what goes on behind the scenes. Flare uses XML, which means you can manipulate, process, and merge content outside of the user interface (UI). The ability to manipulate XML-based Flare content outside the tool provides flexibility you can't get with a tool that uses a proprietary format. For example, you can easily automate your production

---

[1] From MadCap Software: http://www.madcapsoftware.com

process. In Chapter 8, *Document Automation and Batch Files*, we explore a utility to create batch files that build help files or PDFs. We show how to automate content builds and even how to automate creation of the files that define those builds.

You can create deliverables that blend content from different teams. In our workplace, we produce documentation for developers that combines author-created content with content automatically generated from a database. Perhaps you have similar deliverables. Maybe you write configuration manuals that incorporate information extracted from code comments such as permissions tables, or maybe you need to create database documentation that includes both authored content and generated entity-relationship diagrams. By working outside the UI, you can create processes that seamlessly blend this material, and you can automate those processes.

Because the project structure is open and many of the file types are XML-based, working outside of the Flare user interface can be as simple as editing an XML file.

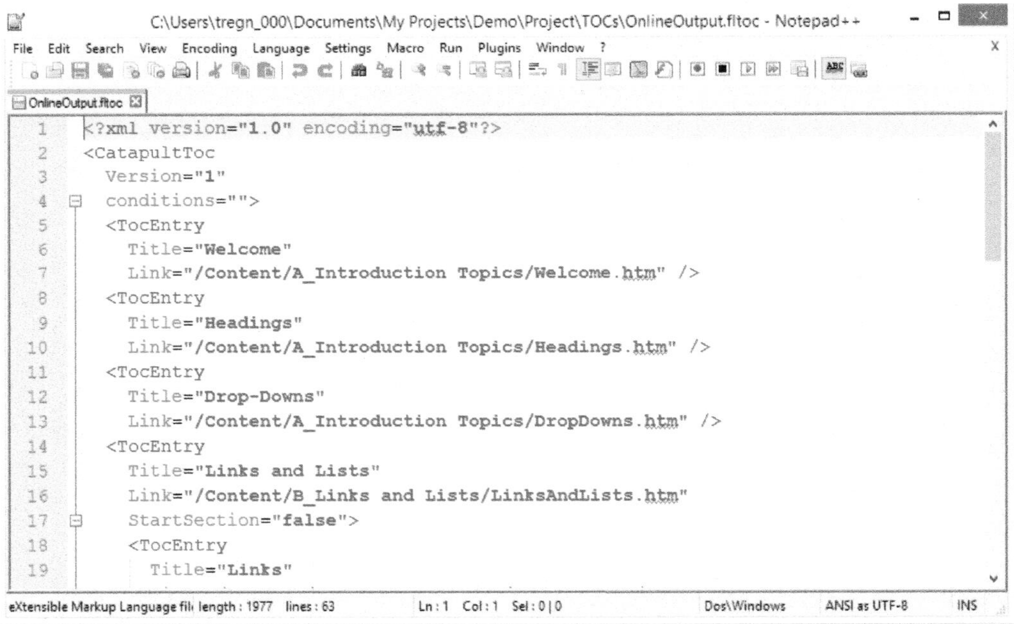

Figure 1.2 – Flare TOC (table of contents) viewed from Notepad++

Audience for this book

This book is for programmers, programmer-technical writers, XML specialists, tech-savvy authors, information architects, and managers – anyone who wants to go beyond what is achievable via the MadCap Flare user interface. Although this book contains code and markup, we don't show you how to code. Our goal is to explain how Flare projects work, describe the files (mostly XML) that make up Flare projects, and show you how those files can be manipulated.

Programmers

If you are responsible for document automation or you need to connect an application to Flare-generated help output, this is the book for you. We describe Flare artifacts and how to create and manipulate those artifacts with technologies such as XSLT, JavaScript, and .NET LINQ to XML. We demonstrate how to programmatically create Flare content, automate documentation builds, connect help outputs to applications, and extend the Flare user interface with plugins.

Technical writers

If you are a technical writer interested in XML or coding, but aren't quite at the point where you can write XSL transformations or automation programs, this book explains some of what is possible and some of the technologies that can be used. This book won't teach you how to write JavaScript or how to code an XSL transformation, but it contains examples you can build on even if you aren't a coder.

For a technical writer with some programming experience, the examples help you begin customizing out-of-the-box Flare outputs. For example, we show you techniques for including JavaScript in your outputs that don't require deep scripting knowledge. Reading the more technical descriptions may help you target areas of interest that you can explore further.

Managers

If you manage a documentation team that uses Flare or plans to use Flare, this book will help you understand what you can do with Flare and how you can go beyond the capabilities of the user interface. You may want to implement some of these ideas to increase your team's efficiency or create new types of output. If you manage a team in a software development environment, you may be able to find a programmer or technically proficient tech writer who can help implement these techniques.

Purpose of this book

This book demonstrates how users can programmatically interact with Flare outside of the user interface. The book shows how content is managed and parsed by Flare, how to work with that content, and the advantages and benefits of doing so. Here are some things to keep in mind:

- The ideas and practices discussed in the book are mainly specific to Flare. However, we delve into larger technical communication subjects such as topic-based authoring and the Darwin Information Typing Architecture (DITA). Most in the technical communication field will be readily familiar with these ideas, but programmers may not be. For that reason we provide background information to explain their role in a Flare environment.
- When discussing XML, we try to be as specific to Flare as possible, but to explain how Flare uses XML, we sometimes need to explain XML usage in general. Unlike DITA and topic-based authoring, XML is likely to be better-known to programmers than to technical writers.

Structure of this book

You can read this book from start to finish, and we considered that when we organized the chapters. However, if you have a particular interest or if you want to skip around, here is what each of the chapters covers:

- Chapter 2, *Flare Projects*, introduces the structure of a Flare project, including the folders and files that make up a Flare project.
- Chapter 3, *Flare, Topic-based Authoring, and DITA*, explores the broader concepts of Flare in the context of topic-based authoring workflows. This chapter is particularly important if you have never used a topic-based or structured authoring methodology.
- Chapter 4, *Flare Topics*, explores topics. We look at topics from the outside (the Flare user interface) and the inside (XML markup). This chapter establishes the basic pattern followed by other chapters in the book: explanation, discussion, and examples.
- Chapter 5, *Flare Tables of Contents*, dives into the primary way for connecting topics together, Tables of Contents (TOCs).
- Chapter 6, *Flare Indexes, Glossaries, and Search*, discusses three mechanisms that enable users of Flare-generated output to find information.
- Chapter 7, *JavaScript in Flare Topics, Master Pages, Snippets, and Skins*, surveys methods for including JavaScript in Flare content.

- Chapter 8, *Document Automation and Batch Files*, shows you how to automate building online and print deliverables.
- Chapter 9, *Connecting Applications to Flare Help Outputs*, shows you how to access Flare help outputs from applications.
- Chapter 10, *Flare Application and Flare Plugins*, provides guidance on how to extend the Flare application user interface with the Flare Plugin API.
- Chapter 11, *Formatting and Pasting Code Samples into Flare*, is a nod to the growing importance of code documentation and is also a follow-up to Chapter 10.
- Chapter 12, *Strategies for Combining Generated and Authored Content*, discusses strategies for combining generated content with authored content.
- Appendix A, *Batch File Generator Utility*, contains the program listing for the Batch File Generator Utility.
- Appendix B, *Element List*, describes some of the more widely used Flare XML elements and attributes.
- Appendix C, *Glossary*, describes the terminology used in this book.

CHAPTER 2
Flare Projects

Flare has a consistent and well-designed user interface. This design creates a workflow for not only creating content, but also for managing Flare projects and outputs, which is at the heart of the single-sourcing nature of the application. For those who are new to Flare, here is some background on the directory structure and files that make up a Flare project.

Project files

When you create a new project (**File → New Project**), the **Start New Project Wizard** (see Figure 2.1) asks you for a name and a folder. Your new project's root folder will go in the folder you specify. So if you enter `NewProject` for the project name and `C:\Projects\` for Project folder, Flare will create a root folder for the project at `C:\Projects\NewProject\` and will place the project file (`NewProject.flprj`) in that folder. To open that project from Flare, select **File → Open** and navigate to `C:\NewProject\NewProject.flprj`.

A project file has a `*.flprj` extension. It is an XML file, but authors won't see that when they use Flare to author content. However, developers who want to manipulate Flare projects programmatically will definitely be concerned with what the XML, JavaScript, CSS, and other component files in a Flare project look like.

MadCap has put a great deal of effort into tailoring the user experience for Flare authors. To do this, they have created file structures that the application expects to see. As a developer, you will be creating programs that manipulate or create Flare files. When you create or alter files in Flare projects, you must make sure that the resulting files conform to what Flare expects. Therefore, you must understand how the application interacts with project files, how it expects those files to be structured, and where it expects those files to be located. We will go into detail about ways to do this in subsequent chapters, but right now it is important to understand that the results of your manipulations must be indistinguishable from those of an author using the Flare interface.

Figure 2.1 – Start a New Project Wizard

Root folder and explorers

With the project open in Flare, you can explore it with many screens, editors, and explorers. The two most important explorers are the Project Organizer and the Content Explorer. These correspond almost directly to folders within the project's root folder. Here are the contents of the root folder immediately after you create a new project:

- **Analyzer:** Folder that contains a database to speed up actions such as Auto-suggest
- **Content:** Folder that corresponds to the Content Explorer in the Flare UI
- **Output:** Folder that contains the output of builds
- **Project:** Folder that corresponds to the Project Organizer in the Flare UI
- **NewProject.flprj:** Project file

Content folder

The Content folder corresponds to the Content Explorer in the Flare UI. The Content folder, not surprisingly, holds items such as topics, snippets (files that hold less-than-a-topic content), master pages, and CSS files. In other words: the content you create.

You can add subfolders to the Content folder. For example, you can place a topic called NewTopic.htm in the Content folder. But you could also place a copy of the same topic in a subfolder called ExtraTopics. Although these two topic files have the same name, they are unique because they have different paths. You have plenty of flexibility in how you structure the content folders.

The templates that come with Flare may contain subfolders; for example, the Resources subfolder appears in most templates. However, you can delete that folder if you want. Unlike some other folders we will encounter later, the Resources folder does not restrict the types of content files it can contain. You can keep master pages, topics, and JavaScript files in the same folder, or you can break them out into separate folders.

If you explore a Content folder outside of the context of the Flare application (for example, by using the Windows (File) Explorer), the folder will look the same as any other folder in the file system. However, if you explore the Content folder with the Content Explorer in Flare, you get tailored, author-centric options.

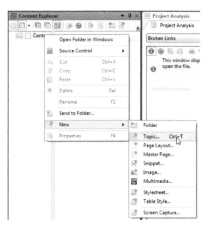

Figure 2.2 – A Right-click menu in Content Explorer

As shown in Figure 2.2, the right-click menu in the Content Explorer lets you add new items such as topics and images. You can also add new items by copying them using Windows Explorer, and those items will also appear in the Content Explorer in Flare. If you do that while the Content Explorer is open, you'll notice a time lag while Flare evaluates what you have placed there.

If you place an `.htm` file in the `Content` folder and attempt to open it from Flare, Flare will validate the file against the Flare schema. If the validation doesn't succeed, Flare will offer to attempt to convert the file. Keep this in mind as you begin generating Flare topics with tools other than Flare. Flare expects files to follow a certain structure. With topics, that structure is defined by a schema (`*.xsd`) file. That's pretty straightforward, but be aware that the algorithms that define Flare's structural expectations for other files may be buried deep in the .NET assemblies that the Flare application uses.

Project folder

The `Project` folder corresponds to the Project Organizer. The Project Organizer manages files such as tables of contents (TOCs), which are hierarchies of topics and other Flare artifacts, and targets, which define the rules for building content and project files into outputs such as PDF and browser-independent WebHelp.

Flare expects each project folder to contain files that correspond to one file type. For example, in Figure 2.3 there is a `TOCs` subfolder under the `Project` folder. When you right-click that folder, you will see **Add Table of Contents** as one of the options, but you won't see an option to add a report or a glossary. To add those, you must right-click a different folder or use the ribbon menu or tool strip. If you go outside of Flare and copy a TOC file (`*.fltoc`) into a folder other than `TOCs`, that file will not appear in the Project Organizer.

You can add subfolders to these folders, but the files you place in them must match the types expected in the parent folder. For example, you may want to organize your target (`*.fltar`) files by target type in separate folders. You can do that, but all the target files must be in some subfolder of `Project\Targets\`.

When you right-click the **Project** node in Project Organizer, you can view a Properties screen for the project. The details for this screen are defined in the `*.flprj` file in the project's root folder.

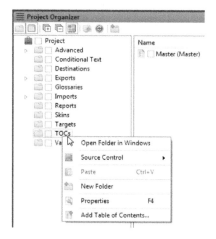

Figure 2.3 – A right-click menu in Project Organizer

Analyzer and Output folders

The Analyzer speeds up actions such as the Spellcheck and Auto-suggest features for variables and snippets. The Analyzer folder contains a database, which Flare uses to store and retrieve information about the content and project files. If you delete the Analyzer folder and database, Flare will recreate them the next time you open the project. The recreated database will be based on the current project structure. The Analyzer database is used for project analysis – it doesn't store content or items that affect Flare output. When Flare scans a project for problems such as broken links, this database plays a part.

The Output folder contains builds of Flare targets, such as PDF and help output. The Output folder also holds temporary files for builds.

You can think of the Analyzer and Output folders as places Flare uses to show information to you, including output builds. And you can think of the Content and Project folders as your folders. The Flare UI imposes structure on these folders, but you create the content.

Conclusion

That is a quick, high-level overview of how a Flare project is structured. The Flare UI is designed to maintain that structure, and other Flare screens and wizards are designed to manage how specific file types are structured. For example, the Topic Editor helps ensure that a topic file adheres to the Flare schema, and the TOC Editor helps ensure that a TOC conforms to the proper hierarchical structure using the XML tags and attributes that Flare expects.

We discuss specific file types in detail in other chapters, and we provide examples of how to manipulate those files programmatically outside of Flare. However, in every case, you must consider what Flare expects for each file type. One of the best ways to understand that is to compare the underlying file, usually in XML format, to how that file appears in a Flare editor.

CHAPTER 3
Flare, Topic-based Authoring, and DITA

Flare is an authoring tool structured around topic-based authoring. In Chapter 4, *Flare Topics*, we describe the XML structure of Flare topics, but here we talk about topic-based authoring in general, topic-based architectures such as DITA, and topic-based authoring in Flare.

Topic-based authoring

Topic-based authoring means creating content in pieces defined as *topics*. That may seem obvious, but there is considerable debate in the tech comm community over what constitutes a topic. Entire books have been written on that subject alone. Technical communicators often differentiate topics by content type, for example, concept, task, or reference. However, for our purposes, a topic can contain whatever you want it to.

No matter how you define a topic, building content in topics enables content reuse and lets you assemble topics into deliverables such as web-based help systems or PDFs. It's also worth noting that the granularity of a topic can be different for the author and information consumers. For example, a consumer would probably call a help file that contains a conceptual overview followed by a procedure one topic, but an author would consider that same help content to be two topics.

DITA

Before we talk much more about topic-based authoring, let's talk about DITA because there are parallels between the DITA architecture and topic-based authoring in Flare. DITA stands for Darwin Information Typing Architecture. DITA is a topic-based XML architecture. DITA provides a mechanism, called specialization, that supports the creation of domain-specific variants that remain compliant with the standard. This capability has helped make DITA a popular standard for technical communicators.

DITA breaks topics into concepts, tasks, and references. Authors assemble topics into deliverables by creating DITA maps. Conceptually, maps can be thought of as tables of contents, but rather than merely showing the hierarchy and organization of information, a DITA map dictates what topics are included and what their hierarchy is within a given deliverable.

A DITA map serves the same function as a Flare TOC file. In fact, when you import a DITA map into Flare, Flare converts it into a Flare TOC file. Because topics exist independently, they have no hierarchy until you impose one. For example, in print, the same topic might be a first-level heading in one book and a second-level heading in another. Many topic-based authoring tools or architectures employ a model similar to DITA maps to organize content.

Originally created by technical writers and developers at IBM, DITA is now an open-source standard managed by the OASIS DITA Technical Committee.[1] Because DITA is a standard, content that follows the standard can be easily merged with DITA content developed by a different group, even if that group has specialized DITA differently. And many open-source tools are available to generate output from DITA content, saving effort for both authors and developers.

Topic-based authoring in Flare

In Flare, topics can be generic, or they can describe a specific type of information. Really, a topic can be whatever you want it to be as long as it adheres to MadCap's schema. Flare lets you associate a topic with an information type but does not require you to do so.

So, is associating topics with a type a good idea?

If you might adopt DITA at some point, it probably is a good idea, and you should consider breaking down topic types into concepts, tasks, and references. This breakdown covers the majority of user assistance deliverables. If you need additional information types for your projects, you can create them in Flare and later use them in DITA. If you anticipate a future move to DITA, you can get a head start by mimicking its conventions in Flare projects. This can make the move easier by helping your team think in terms of the DITA architecture.

When our team moved from structured FrameMaker to a topic-based authoring solution, we initially focused on DITA. Although we decided to go with Flare, we set up our infrastructure to include the DITA information types.

Indeed, the intended use case for Flare classifies topics as concepts, tasks, and references. And you can extend these types as desired. By convention, the value of the `class` attribute on the `<html>` element indicates the topic type. You can ignore that convention if it doesn't fit your

---

[1] https://www.oasis-open.org/committees/dita

workflow, but keep in mind that Flare provides features that use the topic type. For example, Flare can generate a list of concepts based on the `class` attribute.

We go into the detailed structure of Flare topics in Chapter 4 and Flare TOCs in Chapter 5. At a high level, Flare topics are XML files that usually have at least one heading element that corresponds to a heading in XHTML or HTML. Other elements represent paragraphs, proxies, and MadCap controls. Authoring topics in Flare means deciding what information should be kept in each file and deciding how to best arrange those files in a TOC so readers can easily find the topics.

Single source / multichannel

Flare supports single-source multichannel outputs, which means you can have a particular arrangement of topics for one output and another arrangement for another output. If you reuse topics in this way, you must carefully construct them so they make sense regardless of the context you present them in.

That doesn't mean that topics cannot act differently depending on the output. For example, you may have one stylesheet for HTML5 output and another for PDF output. Or, instead of having one stylesheet for each output type, you can maintain a single stylesheet that includes common styles for all outputs along with separate styles for non-print (e.g., CHM) and print (e.g., XPS) outputs, which are applied only to those outputs. For more flexibility, you can link to stylesheet files at the topic and project level.

Conditional content

Flare lets you apply condition tags to include or exclude content depending on the target output type. For example, for an overview paper, you may want your print output to display only a portion of the introductory text but your web-based output to include the complete text. You can accomplish that using condition tags to determine what content will appear (or not appear) in an output. You can also create separate topics for the shorter and longer versions and selectively include those topics based on the output. Flare provides many ways to support conditional content.

Flare and other tools and standards

Many of the same concepts that apply to DITA and Flare also apply to other authoring tools. When you look at Flare, you can see echoes of DITA in the way content is organized and in such features as relationship tables. DITA relationship tables and Flare relationship tables are implemented differently, but the basic idea is the same; they both link related topics together. Relationship tables are useful in a topic-based environment because they aren't coded into topics directly, even though they appear to users as part of a topic. Doing this makes topics easier to reuse. When we changed our authoring environment, we considered DITA, but decided on Flare in part because we were able to mimic the functionality of DITA with relationship tables.

This section compares the Flare topic structure with the structures used by DITA, DocBook, Word, and FrameMaker. We focus primarily on the relative ease of round-tripping content between Flare and each of these tools or structures.

Imports and exports

You can manage your content with Flare projects as the single source for your content. In this frequently used scenario, you perform all or most of your authoring with Flare, and you publish your Flare content to a variety of formats. This is known as multichannel publishing. Flare can can import some or all of your content from other formats. For example, you can import DITA content and publish it to a Flare web-based output. So Flare has the flexibility to be either your main authoring tool or an intermediary help authoring tool.

Imports

Much of the import behavior in Flare can be defined with Flare *import files* (`*.flimp`). These XML files define an import from Word, FrameMaker, DITA, CHM, an uncompiled HTML Help system, or another Flare project. When you run the CHM import wizard, the process creates a new project. However, with all other imports, the Flare import file is kept in a project, and you can trigger the import from either the Import Editor or from a target set to auto-import.

Triggering the import from a target set to auto-import is called the *Auto-Sync* method of integration (also known as *Easy Sync*). You get this action when you choose **Auto-reimport before "Generate Output"** in the Project Import Editor. You can also synchronize outside files with copies in Flare projects with the External Resources feature, which enables you to consistently pull in content that may be updated regularly outside of Flare. Procedure 3.1 shows how to use Auto-Sync.

Flare, Topic-based Authoring, and DITA   17

Procedure 3.1. How to use Auto-Sync

1. Create two Flare projects: one to author topics (`authoring.flprj`) and one to import those topics with Auto-Sync (`importing.flprj`).
2. Create some topics in the authoring project and place those in a TOC.
3. Create a Flare Import File in the importing project: right-click **Project Organizer → Project → Imports** and select **Add Flare Project Import File**. When the **Add File** screen appears, click **Add**.
4. From the Project Import Editor, click **Browse**, which is next to **Source Project**, and select `authoring.flprj`.
5. Ensure the checkbox **Auto-reimport before "Generate Output"** is set.
6. Ensure the **Include Files** list includes topics.
7. Add TOC Files to the list: click **Edit**, then, from the **Import File Filter** screen click **Add**, from the **Import File Filter Designer** screen select **TOC | *.fltoc**, and click **OK** twice.
8. Select **Auto-include linked files**.
9. Save the changes to the Flare Import File.
10. Create a Target File in the importing project. Right-click **Project Organizer → Project → Targets** and select **Add Target**. When the **Add File** screen appears, click **Add**. You can use the defaults.
11. From the Target Editor, ensure **General → Auto-Sync → Disable auto-sync of all import files** is not selected.
12. Save the changes to the Flare Target.
13. Click **Build** from the Target Editor.
14. View the Flare Import from the Project Import Editor. From Imported Files, notice that the list of files includes the TOC and topics from the authoring project.
15. Now make some changes in the authoring project and build the target in the importing project again. Notice the files in the importing project are updated with the changes in the authoring project.

Exports and targets

Flare 10 added a feature to export content. You can also export by creating a build target for the desired output. You can create build targets for a wide range of outputs, including Microsoft Word, DITA, FrameMaker, HTML Help, and PDF. Output types also include systems developed by MadCap, such as HTML5 help and DotNet help.

DITA

If you create a project in Flare, then add a DITA target and generate it, you'll get DITA topics and a DITA map. The output transformation can handle a variety of Flare artifacts. For example, if your target includes a glossary, you'll get a DITA topic for the glossary.

If you then import that DITA content back into Flare, you'll get new Flare artifacts in a different folder. These artifacts include copies of the topics, a new TOC for the DITA map, a relationship table, and a CSS file. This is not a true round trip because the result of importing DITA back into Flare does not return you to the original Flare source. To get a true round trip, you would have to set up a more complex workflow that's outside the scope of this book.

Microsoft Word

Word includes some features that are comparable to Flare. For example, it supports outlining and styles. And rigorous use of styles can provide some structure for Word documents. However, Word documents tend to stand alone as single documents whereas Flare, FrameMaker, and DITA support collections of documents organized into a larger structure. Such collections of documents in Word can be unstable.

Flare can output Word. If you use Flare and need Word output, we recommend you author your content in Flare and output to Word. But maybe you need to import content from Word documents on a regular basis. For example, you may need to pull in content from groups in your company that only use Word. In that case, the best course is to define a well-structured Word template with a minimal number of styles, enforce use of those styles, and build an import for that content. The Auto-Sync method is designed for this kind of workflow.

FrameMaker

When used as a structured authoring tool, FrameMaker supports topic-based authoring. But where structured FrameMaker has books, chapters, formats, elements, and an EDD (Element Definition Document), Flare has topics, TOCs, CSS styles, and MadCap's schema. The most likely workflow between FrameMaker and Flare is a one-time import for authors who are changing their authoring tool from FrameMaker to Flare. However, you can use the Auto-Sync method with FrameMaker, too.

CHAPTER 4
Flare Topics

In this chapter, we delve into the structure of a Flare topic and describe how authors edit topics.

Flare topics overview

A Flare topic contains a single unit of information, independent from any one project. A table of contents (TOC) determines a topic's context and hierarchy in relation to other topics. You can link to a single topic from multiple TOCs, and you can even link to the same topic from multiple places in a single TOC. (see Figure 4.1).[1]

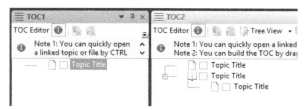

Figure 4.1 – Two TOCs with multiple links to the same topic

Flare topic files can contain tags for headings, paragraphs, reference snippets (reusable bits of content), and images, to name a few; they are not limited to any particular definition of topic. And Flare topic files are not the only XML files in a Flare project. Most Flare files, other than images and other artifacts, use XML. These files all have extensions based on their purpose. For example, Flare TOCs have the extension `.fltoc`, and Flare topics have the extension `.htm`.

Authoring Flare topics

Flare provides two editors for authoring topics. The XML Editor displays a semi-WYSIWYG (What You See Is What You Get) view. We write semi-WYSIWYG because Flare produces output through a build process that is not instantaneous. The end result depends on many factors, including the build target and link resolution. For example, the build process resolves links between topics. Until you run a build, Flare doesn't know exactly how to render links.

---

[1] You can drag-and-drop topics into a TOC, and if you move items, Flare will automatically update links. A Flare wizard takes care of the heavy lifting, offering great flexibility to the authoring workflow.

The preview feature of the XML Editor does a quick build of a topic based on the settings in a given target, but links are rendered with placeholder values.

The XML Editor provides an alternate interface that represents element tags and attributes as blocks. This interface, along with other windows in Flare, provides a middle ground between authoring with a word processor and authoring XML in a plain-text editor.

The Text Editor provides a more native view of the markup for a topic. However, you have the option to colorize markup and use a collapsible display. In this view, Flare still validates your content and displays warnings and errors for bad markup. MadCap recently added a split view that synchronizes the Text and XML Editor views in a single tab (see Figure 4.2).

Figure 4.2 – Topic in the split view

Flare is a tool for multichannel output. Although Flare topics start their lives as XML files, when you build a Flare target, topics morph into PDF sections or HTML5 pages, among other possible outputs. Figure 4.3 shows the same topic rendered in HTML5 and PDF format.

Figure 4.3 – Topic as an HTML5 topic and as a PDF section

Nothing stops you from placing a large amount of loosely related information in a single topic file, separated by multiple headings; a topic is not limited to a single heading. However, the first heading tag in a topic file serves a special purpose. When you drag a topic file into a TOC, Flare creates a TOC entry that links to the topic file through a relative path and uses the first heading as the TOC Label. The first heading in the topic shown in Figure 4.2 is `<h1>Topic Title</h1>`.

Figure 4.4 shows the results of dragging this topic into a TOC. As expected, the TOC Label is "Topic Title." The TOC Label is distinct from the filename. It is also distinct from the topic title, which comes from the `<title>` element that appears in the `<head>` element of XHTML files.

Figure 4.4 – Dragging a topic to a TOC

When you view a topic in the XML Editor in Flare, the topic title is not visible. The XML editor limits access to items within the `<html>` element in the file; you can't see the contents of the `<head>` element. But you can see the title if you view the Properties for a topic using the right-click menu for a topic (see Figure 4.5). The property may be blank. In that case, Flare lets the first heading in the topic stand in for the title.

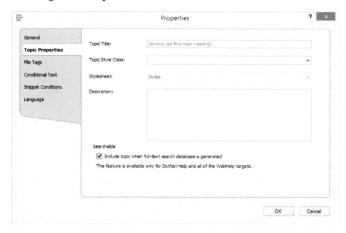

Figure 4.5 – Topic Properties window

Styling Flare topics

Flare uses CSS to guide your selection of elements and `class` attributes for elements in snippets, topics, and master pages. When you highlight text in the XML Editor and select a style from the Styles window, Flare wraps the text with an element and assigns the `class` attribute the value associated with that style, if one has been defined.

The stylesheet determines which elements and styles you can select in the XML Editor. The out-of-the-box CSS stylesheet provides a set of predefined topic style classes: task, concept, reference, and topic. Figure 4.6 shows these styles in the Stylesheet Editor.

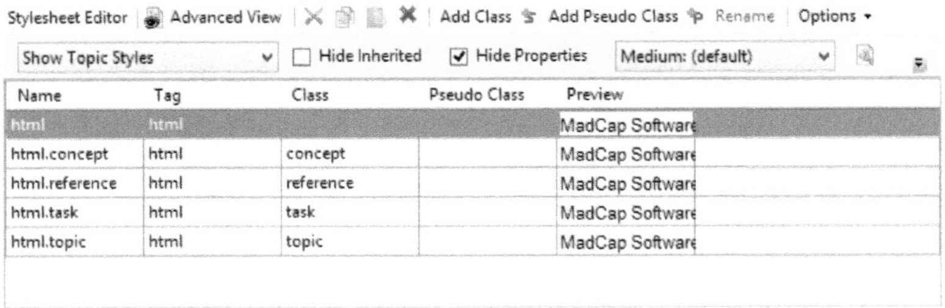

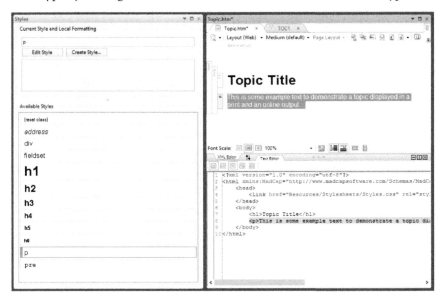

Figure 4.6 – Stylesheet Editor showing task, concept, reference, and topic classes

Using the `class` attribute to define the information type makes it easy to apply distinct styles to each type. And, as with DITA, you can extend the Flare architecture to include additional information types; just assign a new value to the `class` attribute for the new type.

Figure 4.7 – Selecting a style for an element in a topic in the XML Editor

Flare determines which `class` attribute values are appropriate for an element using the CSS stylesheet selected for the project or topic containing that element. By default, Flare uses the stylesheet for the primary target. The XML Editor lets you choose only values that are named in the selected stylesheets (see Figure 4.7). However, you can assign any value to the `class` attribute using the Text Editor or the Attributes screen in Flare.

You can use CSS to associate *master pages* with topics for web output. As shown in Example 4.1, you create a CSS entry that defines the `mc-master-page` style property for the `<html>` element with whatever `class` attribute value you want (in this case, that value is `topic`), and use the chosen `class` attribute on the `<html>` element for that topic.

Example 4.1 – CSS to set `mc-master-page` for the `<html>` element

```
html.topic
{
    mc-master-page: url('../MasterPages/KB_Master_Page.flmsp');
}
```

Using the `class` attribute value `topic` for the `<html>` element of a topic will associate that topic with the master page named `KB_Master_Page.flmsp`.

The Stylesheet Editor displays the properties differently depending on whether you use the Simplified or Advanced view. Most properties are standard CSS properties, but some are MadCap properties, such as the `mc-master-page` style property. For some properties, the Stylesheet Editor provides drop downs with appropriate options. For example, the drop down for the `mc-master-page` style property contains a list of master pages in the project.

The names of the MadCap properties often give an indication about the usage. They usually have names in the form `mc-*`.

There are also some special style names, which usually have the form `MadCap|*` (for example, `MadCap|breadcrumbsProxy`). Flare changes the style names when it builds the output so that they can be used in the CSS definition for a `<div>` wrapper. For example, the properties defined for the style `MadCap|breadcrumbsProxy` will be changed to `div.MCBreadcrumbsBox_0` in the generated output.

Topic structure

In Flare, a topic is an XML file that validates against a schema defined by MadCap Software. This infrastructure provides a great deal of flexibility in working with Flare content. Example 4.2 shows the XML for a simple Flare topic.

Example 4.2 – XML for a simple Flare topic

```
 1  <?xml version="1.0" encoding="utf-8"?>
 2  <html xmlns:MadCap="http://www.madcapsoftware.com/Schemas/MadCap.xsd"
 3         MadCap:lastBlockDepth="2"
 4         MadCap:lastHeight="111"
 5         MadCap:lastWidth="1142">
 6      <head>
 7          <title>Sample Topic Title</title>
 8      </head>
 9      <body>
10          <h1>Sample Topic</h1>
11          <p>This is sample text.</p>
12      </body>
13  </html>
```

Flare topics use XHTML, which is HTML structured using the rules of XML. Flare adds a MadCap-specific schema, which is defined in the xmlns:MadCap attribute (line 2 in Example 4.2). You will find a copy of the schema in the installation folder for the MadCap Flare application. For example, the location for a default installation of Flare version 10 in Windows would be C:\Program Files (x86)\MadCap Software\MadCap Flare V10\Flare.app\Resources\Schemas.

Architecturally, Flare topics combine the structure provided by XML and the familiarity afforded by HTML. Anyone with a little exposure to HTML will understand what is going on in a topic.

In Example 4.2, Line 1 is the XML declaration. While not required by the XML 1.0 standard, Flare expects it to be present, and if you remove it, the XML Editor will flag the file as invalid and offer to convert the document. The XML Editor will not let you change the XML declaration. However, you can alter it outside of the XML Editor, for example, if you need to change the encoding.

Note: If you modify a file outside of Flare, you can check of the quality of your modifications by opening the file in Flare. If the file contains XML that is invalid or doesn't meet Flare's requirements, you will only be able to access that topic using the Text Editor.

Because topics are based on XML and XHTML, the usual characteristics of each technology are exhibited in topic files. An XHTML file has a `<head>` element (line 6 in Example 4.2), which has a child element called `<title>`. When a tabbed browser renders an XHTML file as a web page, the text of the `<title>` element appears in the tab for that page. This should be familiar to anyone who has worked with HTML. You can also change the value of the `<title>` element for a topic in the **Properties** screen by entering the value in the **Topic Title** field on the **Topic Properties** tab.

You can also tell Flare to use the first top-level heading as the Topic Title. In Example 4.2, this is the `<h1>` element on line 10. The result would be `<head><title>`Topic Title`</title></head>` in the generated output. This blurs the distinction between a topic file as a viewed from the Flare interface and as viewed as a pure XHTML file.

In some ways, such as this example, the Flare interface defines a characteristic of the generated topic that is not evident in the underlying source topic markup. There is no attribute or element that tells Flare to use the first heading for the `<title>` element. It just does that when there is no `<title>` element. You will find other linkages between qualities of a topic file and what is exposed on the **Properties** screen. For example, there is a direct correspondence between the file name and the **Name** field of the **General** tab.

The point here is to understand how Flare exposes XHTML documents in its user interface and what Flare does with those topics when it creates an output. Once you have that understanding, you can feel comfortable about how to create and manipulate files outside of the Flare interface.

That is what makes the open XML project structure of Flare so powerful. There is a user interface with an acceptable learning curve that you can use to manage the project structure. But if you want to push the envelope, the project structure does not restrict you. If you want to create a topic with Notepad.exe or VIM or create 10,000 topics using a homegrown utility, go ahead; there is nothing to stop you.

Contents of a Flare topic

A Flare topic can contain a wide variety of content. Here are some of the more important types:

XHTML element tags

Topics, as well as snippets and master pages, can contain standard XHTML tags such as `<p>`, `<h1>`, and `<div>`. XHTML rules apply.

MadCap element tags

Snippets, topics, and master pages can also contain proprietary tags prefixed with the MadCap namespace, which is defined in the MadCap schema: `MadCap.xsd`. These tags define items such as proxies for mini-TOCs, breadcrumbs, topics themselves, and placeholders to display relationship tables. Other MadCap tags define UI controls for web-based output such as Togglers and Popups. You can find a list of commonly used MadCap tags in Appendix B, *Element List*.

Links, cross references, and bookmarks

Links are defined using `<a>` tags, as in HTML. Cross references use `<MadCap:xref>` tags. The `<MadCap:xref>` tag, used in conjunction with stylesheets, enables special behavior in outputs, such as inserting predefined text into your output. For example, you could add text such as "on page x" to a cross reference in print outputs. Bookmarks also use the `<a>` tag and `name` attribute. This works like HTML, where the name or id identifies the bookmark and the `<a>` element can point to the bookmark by referencing the `name` or `id` attribute value. Flare also interprets bookmarks for non-web outputs such as PDF.

Snippets

Snippets are similar to topics, but they are not referenced individually from a TOC. Snippets can be embedded inside topics, master pages, or other snippets. Snippet files resemble topic files, but they represent a fragment of content that can be reused in many places. This chapter doesn't focus on snippets, but you can use the same techniques described here for topics to create and manipulate snippets using a program.

Snippets are inserted into a topic using the `<MadCap:snippetBlock>` element. This element takes the `src` attribute, which contains the relative path to the snippet (see line 10 in Example 4.3). The entire snippet is pulled in, replacing the `<MadCap:snippetBlock>` element. The snippet must be valid XML and must also be valid in the context it's pulled into. Output processing is done after snippets have been processed. Snippets that just contain text, with no elements or tags, can be embedded using the `<MadCap:snippetText>` element.

Example 4.3 – Code to insert a snippet in a topic

```
1  <?xml version="1.0" encoding="utf-8"?>
2  <html xmlns:MadCap="http://www.madcapsoftware.com/Schemas/MadCap.xsd">
3    <head>
4      <link href="Resources/Stylesheets/Styles.css"
5            rel="stylesheet" type="text/css" />
6    </head>
7    <body>
8      <h1>Sample Topic</h1>
9      <p>This is sample text.</p>
10     <MadCap:snippetBlock src="Resources/Snippets/sample-snippet.htm" />
11   </body>
12 </html>
```

Conditions

Conditions can be applied at many levels in Flare projects. You can apply conditions to most XML elements, including the following: topics, snippets, paragraphs, and TOCs. This allows you to apply conditions at nearly any level of granularity. When you generate an output, you can include or exclude content based on the value of any condition defined on that element.

Conditions are implemented using the `MadCap:conditions` attribute. Line 9 in Example 4.4 is an example of a condition applied to a paragraph.

When an element contains a condition, and that condition is not excluded from the output, the condition passes through to the output as a `MadCap:conditions` attribute. This attribute can be used for dynamic filtering using JavaScript.

Example 4.4 – Topic with a condition applied to an element

```
1  <?xml version="1.0" encoding="utf-8"?>
2  <html xmlns:MadCap="http://www.madcapsoftware.com/Schemas/MadCap.xsd">
3    <head>
4      <link href="Resources/Stylesheets/Styles.css"
5            rel="stylesheet" type="text/css" />
6    </head>
7    <body>
8      <h1>Sample Topic</h1>
9      <p MadCap:conditions="Default.PrintOnly">This is sample text.</p>
10   </body>
11 </html>
```

Variables

Variables can be used in snippets, topics, and master pages. Flare uses the `<MadCap:variable>` element with the `name` attribute to reference a variable. You define the value for each variable in your target file. Flare 10 lets you define multiple values for a variable. The first value is the default, but you can override that value.

Proxies

Proxies are placeholders for special behavior in generated output. Proxies are used for such functions as breadcrumbs, lists of related links, and mini-TOCs. Proxies are also used to indicate the placement of topic content on a master page.

Web outputs apply a master page to display a web page and place the topic content wherever the topic proxy tag (`<MadCap:bodyProxy/>`) is located. You can also use proxies for navigation controls in web-based outputs. Although you can define navigation controls on the skin for web-based outputs, you can also place controls on master pages and in topics.[2]

Figure 4.8 shows breadcrumbs generated by a proxy in HTML5 output. Example 4.5 shows how proxy markup can be embedded in a source topic. Example 4.6 shows an HTML5 topic generated from the topic in Example 4.5, including the output from the proxies.

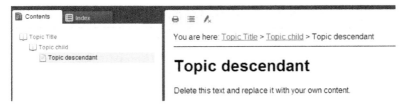

Figure 4.8 – Breadcrumbs in HTML5 output

---

[2] HTML5 output does not support as many toolbar options in topics as its predecessors, such as MadCap's WebHelp. A better approach for HTML5 is to use the skin to create navigation controls.

Example 4.5 – Examples of proxy markup in a source topic

```
1  <?xml version="1.0" encoding="utf-8"?>
2  <html xmlns:MadCap="http://www.madcapsoftware.com/Schemas/MadCap.xsd"
3         MadCap:lastBlockDepth="2"
4         MadCap:lastHeight="273"
5         MadCap:lastWidth="1116">
6   <head></head>
7   <body>
8     <h1>Topic With Proxy Examples</h1>
9     <p>Breadcrumbs proxy:</p>
10    <MadCap:breadcrumbsProxy />
11
12    <p>TOC proxy:</p>
13    <MadCap:tocProxy />
14
15    <p>Mini-TOC proxy:</p>
16    <MadCap:miniTocProxy />
17  </body>
18 </html>
```

Example 4.6 – Generated proxy markup in an HTML5 topic

```
1  <!DOCTYPE html>
2  <html xmlns:MadCap="http://www.madcapsoftware.com/Schemas/MadCap.xsd"
3         lang="en-us" xml:lang="en-us" data-mc-search-type="Stem"
4         data-mc-help-system-file-name="Default.xml"
5         data-mc-path-to-help-system="../" data-mc-target-type="WebHelp2"
6         data-mc-runtime-file-type="Topic" data-mc-preload-images="false"
7         data-mc-in-preview-mode="false"data-mc-toc-path="">
8   <head>
9     <meta charset="utf-8" />
10    <meta http-equiv="Content-Type"
11          content="text/html; charset=utf-8" />
12    <link href="../Skins/Default/Stylesheets/TextEffects.css"
13          rel="stylesheet" type="text/css" />
14    <link href="../Skins/Default/Stylesheets/Topic.css"
15          rel="stylesheet" type="text/css" />
16    <title>Topic With Proxy Examples</title>
17    <link href="Resources/Stylesheets/Styles.css"
18          rel="stylesheet" type="text/css" />
19    <script src="../Resources/Scripts/jquery.min.js"
20            type="text/javascript"></script>
21    <script src="../Resources/Scripts/plugins.min.js"
22            type="text/javascript"></script>
23    <script src="../Resources/Scripts/require.min.js"
24            type="text/javascript"></script>
25    <script src="../Resources/Scripts/require.config.js"
```

```
26              type="text/javascript"></script>
27     <script src="../Resources/Scripts/MadCapAll.js"
28              type="text/javascript"></script>
29   </head>
30   <body>
31     <h1>Topic With Proxy Examples</h1>
32     <p>Breadcrumbs proxy:</p>
33     <div class="MCBreadcrumbsBox_0">
34       <span class="MCBreadcrumbsPrefix">You are here: </span>
35       <span class="MCBreadcrumbs">Topic With Proxy Examples</span>
36     </div>
37     <p>TOC proxy:</p>
38     <div class="MCTocProxy_0">
39       <table style="width: 100%;"
40              cellspacing="0" cellpadding="0" class="GenTOCTable1">
41         <col style="width: 0pt;" />
42         <col />
43         <col style="width: 10pt;" />
44         <tr>
45           <td class="mcReset" />
46           <td class="GenTOCText1">
47             <a class="GenTOCText1" href="Topic.htm">Topic</a></td>
48           <td class="GenTOCPageText1">
49             <MadCap:xref style="mc-format: '{page}';"
50               class="TOCPageNumber"
51               data-mc-xref-target=""> <![CDATA[ ]]></MadCap:xref>
52           </td>
53         </tr>
54       </table>
55     </div>
56     <p>Mini-TOC proxy:</p>
57     <div class="MCMiniTocBox_0">
58       <p class="MiniTOC1_0">
59         <a class="MiniTOC1" href="Topic.htm">Topic</a>
60       </p>
61     </div>
62   </body>
63 </html>
```

Topic output

With a default build for web-based outputs, Flare creates transformed versions of all topics in the folder and file structure. You can choose to omit unused topics from web-based outputs with the **Exclude content not linked directly or indirectly to the target** option, and you can use condition tags to control whether topics are exposed in the navigation.

HTML5 and web outputs
Depending on the output type, topics in web-based outputs are rendered in HTML or XHTML. Placeholder elements, such as variables and proxies, are replaced by variable values and proxy content respectively. For the most part, web-based outputs follow a tri-pane help-system pattern.[3]

PDF and print outputs
In print outputs, topics usually correspond to sections. For example, in Word output from Flare, topic headers may correspond to heading styles followed by paragraph content. However, this mapping is only as strict as the structure of the Flare topic files.

XHTML
Topics correspond to sections delineated by headings. An XHTML Book Target produces a single file web page.

DITA
Flare topics become DITA topics.

Import behavior
You define import behavior in an XML file (`*.flimp*`). Flare import files can define import behavior for Word, FrameMaker, DITA (`*.flimpdita`), and Flare (`*.flimpfl`) content. With Word, sections in the Word document map to Flare topics based on specified styles, typically headings. FrameMaker imports rely on unstructured tags and ignore elements in structured FrameMaker documents. DITA topics and Flare topics map to Flare topics.

Flare Project Imports do not require mapping anything since the files are native. But stylesheets can be imported as part of the Flare Project Import process. Markup for a Flare Project Import File (`*.flimpfl`) is shown in Example 4.7. The root element for a Flare Project Import File is `<CatapultProjectImport>`. Attributes include `AutoSync`, `AutoExcludeNonTaggedFiles`, `IncludeLinkedFiles`, `IncludePattern`, `DeleteStale`, and `ConditionTagExpression`.

---

[3] An exception is the "frameless" HTML5 help introduced in version 11.

Example 4.7 – Flare Project Import File markup

```xml
<?xml version="1.0" encoding="utf-8"?>
<CatapultProjectImport
  AutoSync="true"
  IncludePattern="*.htm;*.html;*.fltoc"
  IncludeLinkedFiles="true"
  LastImport="2015-02-24T10:24:54.2452913-05:00">
  <Files>
    <Url
      Source="file:///C:/Projects/ex/ex.flprj"
      AbsoluteSource="file:///C:/Projects/ex/ex.flprj" />
  </Files>
  <ExcludedFiles />
  <CollectedSourceFiles>
    <Url
      Source="../../../ex/Content/Topic.htm"
      AbsoluteSource="file:///C:/Projects/ex/Content/Topic.htm" />
    <Url
      Source="../../../ex/Project/TOCs/Master.fltoc"
      AbsoluteSource="file:///C:/Projects/ex/Project/TOCs/Master.fltoc" />
  </CollectedSourceFiles>
  <SkippedSourceFiles />
  <GeneratedFiles>
    <Url
      Source="../../Content/Topic.htm"
      TimeStamp="6/3/2013 1:48:14 PM" />
    <Url
      Source="../TOCs/Master.fltoc"
      TimeStamp="6/3/2013 1:48:14 PM" />
  </GeneratedFiles>
</CatapultProjectImport>
```

You can also use a feature called External Resources to synchronize files from outside the project to a copy in the project.

Manipulating topic files

During the authoring process, you can change a topic at several points. Most of the time, you will change topics when creating or editing them. However, you can also manipulate topic files in other ways, including the following:

- **Editing topic templates:** When you create a topic, you base it on an existing topic file, a template. If you change a template, you affect all topics based on that template.

- **Editing the Flare project outside of Flare:** You can manually open and edit topics in text and XML editors outside of Flare, and you can also write programs that manipulate topics.
- **Editing the output:** Another interesting possibility is to edit the output, either manually or with a program. However, keep in mind that you have to repeat your edits every time you generate the output, since any changes you make will be overwritten.
- **Adding dynamic behavior:** You can insert scripts, usually JavaScript, to create dynamic behavior in web-based outputs. For example, you can add a custom filter drop down and script to a topic. A web-based topic can do pretty much anything a web page can do.
- **Manipulating the iframe:** In Flare's HTML5 tri-pane output, the skin displays a topic in an `<iframe>` element (see Figure 4.9). An iframe includes the content of another page. It is a kind of transclusion. When you click a link to a topic, such as a cross-reference or a node on the **Contents** tab, the iframe is refreshed with the selected topic. The iframe provides another intervention point for scripting and dynamic content. For example, you can create an interface that lets the reader use tabs to switch between the out-of-the-box Flare topic and content that you script. This functionality isn't specific to Flare; you can do this with any web page.

Figure 4.9 – HTML5 output for a topic

Topic code samples

Modifying topics programmatically

Let's consider a common text operation on topics: find and replace. When you scan text to find a word with your eyes, you are performing a personal heuristic search. However, as humans, we sometimes overlook things, and we can't parse a text document as quickly as a computer. Search algorithms come in handy for manipulating text and markup.

Flare has find-and-replace functionality. But there are a couple of catches, especially if you want to automate the process. First, you have to use the Flare interface to use its find-and-replace feature; there is no API. Second, this feature has limitations when used for elements and attributes.

The most basic find-and-replace capability matches text exactly. For example, if you search for "Cat," you will find "Cat" but not "cat" or "dog." Some systems let you choose whether a search is case-sensitive or not. A non-case-sensitive search for "Cat" would match "Cat" and "cat" but, of course, not "dog." More advanced find-and-replace capabilities allow you to use wildcard characters and maybe even regular expressions. For example, depending on the wildcard syntax, "*at" will match "cat" and "bat."

Regular expression languages fall under a class of languages in formal language theory called, not surprisingly, regular. A full explanation of regular expressions is outside the scope of this book, but if you are going to do much programming, you will find regular expressions to be very useful.

However, as useful as regular expression languages are, HTML and XML cannot be fully parsed using regular expressions. While you can do some useful work with regular expression languages, they fail when you try to handle tasks such as matching the closing tag of a particular XML element that is nested inside elements of the same type and where the content is otherwise unknown. We need different tools to effectively manipulate XML content.[4]

XSLT examples

If we want to find and replace elements, we can appeal to a higher order of languages that can parse XML and HTML. One such implementation is XSLT (eXtensible Stylesheet Language Transformations). Transforming an XML document is a major function of XSLT. One simple transformation is to change every instance of a particular element in an XML document into a different element in the transformed document. The XSLT in Example 4.8 changes every `<h1>` element to an `<h2>` element. This may come in handy if your Flare workflow requires specific heading levels at the beginning of every topic.

---

[4] It's worth noting two things: First, it is possible to identify starting and closing tags of elements with regular expressions when you know that the element does not contain other elements. Second, some languages called regular expressions can identify the starting and closing tags for an element with nested elements, but these languages are not actually regular languages; they usually implement a behavior called backtracking, which is not a capability of regular languages.

Example 4.8 – XSLT that transforms every `<h1>` element into an `<h2>` element

```
1  <?xml version="1.0" encoding="utf-8"?>
2  <xsl:stylesheet
3      xmlns:xsl="http://www.w3.org/1999/XSL/Transform"
4      xmlns:msxsl="urn:schemas-microsoft-com:xslt"
5      version="1.0"
6      exclude-result-prefixes="msxsl">
7
8    <xsl:output method="xml" indent="yes"/>
9    <xsl:template match="node( ) | @*">
10     <xsl:copy>
11       <xsl:apply-templates select="@* | node( )"/>
12     </xsl:copy>
13   </xsl:template>
14   <xsl:template match="h1">
15     <h2>
16       <xsl:apply-templates select="@* | node( )"/>
17     </h2>
18   </xsl:template>
19 </xsl:stylesheet>
```

An XSL transform treats an XML document like a tree. It traverses the tree and looks for nodes (elements or attributes) that match certain templates. Lines 14–18 in Example 4.8 are a template that matches each `<h1>` element and replaces it with an `<h2>` element. Lines 9–13 match everything else, copying anything that isn't an `<h1>` element to the new document unchanged.[5] The result is a document that is identical to the input, except every `<h1>` element has been changed to `<h2>` (see Example 4.10).

To run an XSL transform, you need an XSL processor. There are many to choose from. For this book, we chose Microsoft's free MSXSL utility (`msxsl.exe`). We downloaded a copy of `msxsl.exe`[6] from MSDN and placed it in a folder called `C:\msxsl`. We then created a batch file to hold the command for the transformation. Finally we placed a Flare topic file with an `<h1>` element and the XSLT file in the same folder as `msxsl.exe`. Example 4.9 shows the batch file.

---

[5] This template is known as the "identify transform." It can be found in many XSL transforms.

[6] http://www.microsoft.com/en-us/download/details.aspx?id=21714

Example 4.9 – Batch file to run `msxsl.exe`

```
cd c:\msxsl
msxsl "h1-example.htm" "change-h1-to-h2.xsl" -o "updated-example.htm"
pause
```

Once run, the output appears as shown in Example 4.10.

Example 4.10 – Transformed Flare topic markup (output from Example 4.8)

```
<?xml version="1.0" encoding="UTF-8"?>
<html MadCap:lastBlockDepth="2"
      MadCap:lastHeight="153" MadCap:lastWidth="1176"
      xmlns:MadCap="http://www.madcapsoftware.com/Schemas/MadCap.xsd">
  <head>
    <link href="Resources/Stylesheets/Styles.css"
          rel="stylesheet" type="text/css">
    </link>
  </head>
  <body>
    <h2 class="heading">Topic Title</h2>
    <p>Delete this text and replace it with your own content.</p>
    <MadCap:indexProxy style="mc-index-headings: true;">
    </MadCap:indexProxy>
  </body>
</html>
```

Example 4.11 matches an attribute instead of an element. It finds every element that has a `class` attribute with the value `main-class` and changes the value to `other-class`. The "@" character in the match on line 13 identifies the thing being matched as an attribute rather than an element. Then, that template (lines 13–24) checks the value of the `class` attribute, and if the value is `main-class`, it changes the value to `other-class`. Then, as with Example 4.8, it copies everything else to the output unchanged.

Using XSLT to transform XML is a natural choice because XSLT was designed specifically to deal with XML. However some find the syntax to be cumbersome and hard to work with.

Example 4.11 – XSLT that finds a particular attribute-value pair and replaces the value

```
1  <?xml version="1.0" encoding="utf-8"?>
2  <xsl:stylesheet
3      xmlns:xsl="http://www.w3.org/1999/XSL/Transform"
4      xmlns:msxsl="urn:schemas-microsoft-com:xslt"
5      version="1.0"
6      exclude-result-prefixes="msxsl">
7    <xsl:output method="xml" indent="yes"/>
8    <xsl:template match="node( ) | @*">
9      <xsl:copy>
10       <xsl:apply-templates select="@* | node( )"/>
11     </xsl:copy>
12   </xsl:template>
13   <xsl:template match="@class">
14     <xsl:attribute name="class">
15       <xsl:choose>
16         <xsl:when test=". = 'main-class'">
17           <xsl:text>other-class</xsl:text>
18         </xsl:when>
19         <xsl:otherwise>
20           <xsl:value-of select="." />
21         </xsl:otherwise>
22       </xsl:choose>
23     </xsl:attribute>
24   </xsl:template>
25 </xsl:stylesheet>
```

.NET examples

There is more than one way to achieve the objectives of the previous examples. In Chapter 10 we explore Flare plugins, which are extensions to the Flare application. Flare plugins are .NET assemblies. If you want to expose a find-and-replace tool in the Flare application, consider using .NET code so that you can expose that tool as a Flare plugin.

You can work with XSLT in .NET. But there are other XML manipulation libraries available as well. One is called *LINQ to XML*. LINQ to XML translates XML into objects and then back to XML at runtime. For example, if you want to create a Flare plugin with C#, a .NET language, you can load topic files into your plugin using LINQ to XML and manipulate the object model for the XML document programmatically. Then you can save the XML document to a new location or replace the document you originally loaded into the plugin. Example 4.12 shows some LINQ to XML code in C# that loads a Flare topic, changes the title, and saves the document.

Example 4.12 – C# console application to change a topic title

```
using System;
using System.Collections.Generic;
using System.Linq;
using System.Text;
using System.Xml.Linq;

class Program
{
    static void Main(string[] args)
    {
        XDocument FlareTopic = XDocument.Load(args[0]);
        if (FlareTopic.Root.Element("head").Element("title") != null)
        {
            FlareTopic.Root.Element("head").Element("title").Value = args[1];
        }
        else {
            XElement title = new XElement("title");
            title.Value = args[1];
            FlareTopic.Root.Element("head").Add(title);
        }
        FlareTopic.Save(args[0]);
    }
}
```

In Example 4.12, the code would be maintained in a C# class in a Visual C# console application. The application expects two string arguments to be passed from the command that executes the application. The array items `args[0]` and `args[1]` correspond to those arguments. A command to execute the program would look like this:

```
ConsoleApplication1.exe "C:\blank\Content\Topic.htm" "New Title"
```

Another nice feature of LINQ to XML is the Visual Basic .NET implementation. In the Visual Basic editor in Visual Studio, you can edit XML literals in-line with Visual Basic code. A programmer familiar with Visual Basic can work in tandem with a technical writer who knows XML. Example 4.13 is a Visual Basic .NET sample that creates a Flare topic file and appends a table to the end of the topic. Notice the XML is easy to see in Example 4.13.

An information architect may want to do this kind of preparation before asking team members to add reference content to a large number of topics. Imagine hundreds of topics that each describe the purpose of a form in an application but do not describe the purpose for each field. Adding tables to each topic provides a place holder for those field descriptions.

Example 4.13 – Visual Basic .NET console application that appends a table to a topic

```
Module Module1
  Sub Main()
    Dim FlareTopic As XDocument = _
      <?xml version="1.0" encoding="utf-8"?>
      <html xmlns:MadCap="http://www.madcapsoftware.com/Schemas/MadCap.xsd">
        <head>
          <link href="Resources/Stylesheets/Styles.css"
                rel="stylesheet" type="text/css"/>
        </head>
        <body>
          <h1>Topic Title</h1>
          <p MadCap:conditions="Default.PrintOnly">Text</p>
        </body>
      </html>

    FlareTopic.Root.Element("body").Add(
      <table>
        <tr>
          <td>first row, first col</td>
          <td>first row, second col</td>
        </tr>
        <tr>
           <td>second row, first col</td>
           <td>second row, second col</td>
        </tr>
      </table>)

    MsgBox(FlareTopic.ToString)
  End Sub
End Module
```

Figure 4.10 shows what the application from Example 4.13 looks like when it is running.

Although LINQ to XML provides some nifty run-time processing, the DOM (Document Object Model) manipulation features introduced in the Flare plugin API with version 10 use XmlDocument from System.Xml. You can still use LINQ to XML in plugins, but you should familiarize yourself with both System.Xml and System.Xml.Linq. Plugins are discussed in depth in Chapter 10, *Flare Application and Flare Plugins*.

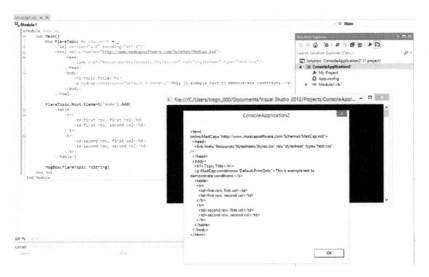

Figure 4.10 – Visual Basic .NET console application running

JavaScript example

Technologies such as XSLT and .NET are useful for manipulating a topic file in a source project. But what if you want to change something in a generated topic? That is, a topic in a web-based help output such as Flare's HTML5 output. Since the output is HTML rather than XML, the best choice is a scripting language designed to manipulate HTML, such as JavaScript.

You can add scripts to Flare topics, master pages, toolbars, and skins, and Flare will include them in the generated output. This helps you manage website behavior. You can do anything that can be done with HTML and JavaScript.

Chapter 7 describes how to add JavaScript, but to whet your appetite, here is some sample JavaScript using jQuery that underlines elements with the `class` attribute value `ProductsFlare`, which MadCap uses to make the product name stand out. Using the Google Chrome console, we will directly insert this JavaScript example.

The page you will be working with is the Flare welcome page. Normally, you would go to the top of the help system, but in this case, we will bypass the iframe and go directly to the content for the welcome page.

Example 4.14 – JavaScript fragment to underline content with `class="ProductsFlare"`

```
$('.ProductsFlare').each(function() {
    $(this).css('text-decoration', 'underline');
});
```

Go to the welcome page,[7] then open the Google Console (**View → Developer → JavaScript Console**), paste in the JavaScript from Example 4.14, and type **return**. Figure 4.11 shows the result.

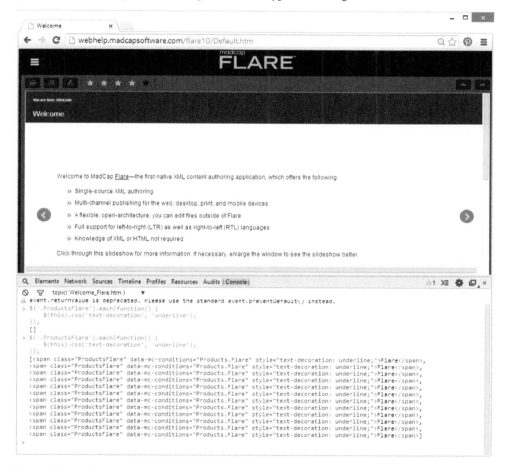

Figure 4.11 – Example underline script in action

---

[7] http://webhelp.madcapsoftware.com/flare10/Content/Introduction_Topics/Welcome_Flare.htm

The console lists more elements with the `ProductsFlare` class value than are visible, but that is because those elements appear in other places in Flare's slider UI (which was introduced with the version 10 release). If you use the left and right arrows on the page, you will find other underlined instances of the term Flare.

In Chapter 7, you'll learn how to create a button in HTML5 output to run other scripts like this. After you've learned how to connect scripts to online outputs, you are only limited by the potential of HTML and scripting.

CHAPTER 5
Flare Tables of Contents

TOC structure

Flare TOCs determine not only which topics are included in a deliverable, but also the hierarchy of those topics in the deliverable. TOCs are the glue that bonds topics into cohesive deliverables. Flare TOCs are maintained in an XML file with the `.fltoc` extension and are stored in the `Project\TOCs` folder.

Let's explore the XML structure of a Flare TOC. Example 5.1 shows the XML for a TOC in a Flare project that has a single Flare topic.[1]

Example 5.1 – XML for a TOC with a single link

```
1  <?xml version="1.0" encoding="utf-8"?>
2  <CatapultToc
3     Version="1">
4     <TocEntry
5        Title="Sample Topic"
6        Link="/Content/sample-topic.htm" />
7  </CatapultToc>
```

The element `<CatapultToc>` (line 2 in Example 5.1) is the root element of a Flare TOC file (`*.fltoc`). The root element can contain one or more `<TocEntry>` elements. You can think of `<TocEntry>` elements as nodes in the TOC tree. A `<TocEntry>` element minimally contains a `Title` attribute. In outputs, the value of the `Title` attribute usually becomes the text of the label for that entry or node.

The `<TocEntry>` element can have a `Link` attribute, which can point to one of several things, including a topic, other TOCs, or a help system. The `Link` attribute specifies a path that is relative to the Project folder. The `AbsoluteLink` attribute lets you specify an absolute path.

---

[1] Most of the Flare XML files, including this one, do not name a schema in the file. However, schemas do exist for these files. You can find them in this folder: C:\Program Files\MadCap Software\MadCap Flare V4\Flare.app\Resources\Schemas

The `<TocEntry>` element in Example 5.1 (lines 4-6) points to a topic. However, a Flare TOC can contain nodes for more than just topics. Example 5.2 shows the XML for a TOC that points both to another Flare TOC (lines 7-9) and to a Flare project (lines 10-13).

Example 5.2 – XML for a TOC with a variety of node types

```
 1  <?xml version="1.0" encoding="utf-8"?>
 2  <CatapultToc
 3    Version="1">
 4    <TocEntry
 5      Title="Topic"
 6      Link="/Content/Topic.htm"></TocEntry>
 7    <TocEntry
 8      Title="OtherTOC"
 9      Link="/Project/TOCs/OtherTOC.fltoc"></TocEntry>
10    <TocEntry
11      Title="HTML5Target"
12      Link="../../../Example/Example.flprj#HTML5"
13      AbsoluteLink="file:///C:/Example/Example.flprj#HTML5"></TocEntry>
14    <TocEntry
15      Title="No link" />
16  </CatapultToc>
```

Nested TOCs

TOCs can reference other TOCs. Nesting helps you chunk TOCs into more manageable sizes for large information systems. Authors can drill down through nested TOCs in the TOC Editor, which opens another TOC Editor for each nested TOC. Lines 7-9 in Example 5.2 show a `<TocEntry>` for a nested TOC. The XML for a nested TOC is identical to the XML for any TOC. You can see this in Example 5.3 and Example 5.4.

Example 5.3 – XML for a TOC with a sub-TOC

```
<?xml version="1.0" encoding="utf-8"?>
<CatapultToc
  Version="1">
  <TocEntry
    Title="ChildTOC               "
    Link="/Project/TOCs/ChildTOC.fltoc" />
</CatapultToc>
```

Example 5.4 – Sub-TOC or a standalone TOC (`ChildTOC.fltoc`)

```
<?xml version="1.0" encoding="utf-8"?>
<CatapultToc
  Version="1">
  <TocEntry
    Title="Topic Title"
    Link="/Content/Topic.htm" />
</CatapultToc>
```

Other TOC node types

In addition to topics and other TOCs, a Flare TOC can link to other information types. Here are some TOC entries for those other types:

Browse sequences

Flare TOCs have a sibling called a *browse sequence*. The difference between a browse sequence and a TOC is that while a TOC presents content in a hierarchical order, topics can be read in whatever order the user prefers; browse sequences define a specific reading order. You use browse sequences to give readers alternative paths through your content. Outputs that use browse sequences typically provide navigation aids that help users remain oriented as they browse forward and backward through the sequence. Here is a `<TocEntry>` for a browse sequence:

```
<TocEntry
    Title="BrowseSequence"
    Link="/Project/Advanced/Browse Sequences/BrowseSequence.flbrs" />
```

We focus on TOCs in this book, but because the markup for a browse sequence is the same as markup for a TOC, you can manipulate them in the same way you manipulate TOCs. Browse sequences are maintained in the **Project → Advanced → Browse Sequences** folder in a Flare project instead of the **Project → TOCs** folder.

Projects and targets

When you specify a project and target in a `<TocEntry>` and build the target, Flare will also build that subtarget and include it in the output for any target that uses that TOC. For this to work, the target type for the parent TOC and the child project and target must be the same. For example, in HTML5 targets, the child target output is placed in a subsystems folder in the parent target output. This subsystem can be accessed independently. In the `Link` attribute of a `<TocEntry>` for a target, there is a path to the Flare project (`*.flprj`) file followed by bookmark that indicates the target (see Example 5.5).

Example 5.5 – TocEntry with a bookmark indicating an HTML5 target

```
<TocEntry
    Title="HTML5Target"
    Link="../../../Example/Example.flprj#HTML5"
    AbsoluteLink="file:///C:/Example/Example.flprj#HTML5">
</TocEntry>
```

HTML Help
You can nest an HTML Help system, but you can only use HTML Help targets in TOCs that are used by HTML Help or CHM targets.

External help
As with project and target entries, the generated output for external help includes the help system as a subsystem. But in this case, the reference is to an already generated system. The term *external help* refers to an already-generated help system of the same output type as the target. Therefore, the workflow for external help must generate (or re-generate) the referenced system first.

Mimic
Mimic is the application for making video screencasts in the MadCap MadPak suite. You can place references to Mimic movies in Flare TOCs.

Import behavior for TOC hierarchies

Imports from Word files usually create a hierarchy defined by a mapping of Word styles to Flare tags. The structure is flexible, depending on the nesting level of the Flare tags (h1, h2, p, etc.).

Imports from FrameMaker files also enable a mapping of FrameMaker tags to Flare tags. The import process creates a mapping, but custom mapping can be defined on the import file with the Import Editor.

Imports from other Flare projects and DITA projects are more straightforward. The hierarchy maps directly from the Flare TOCs or the DITA maps.

Manipulating TOCs

TOCs can be manipulated or generated programmatically. Sometimes it makes sense to generate a TOC when you create topics. Other times it makes sense to generate a TOC based on a set of existing topics.

The latter approach is especially applicable when generating references. An API reference, for example, may have thousands of topics generated by a utility. Or your workflow may involve generating many topics, but only including a subset of those in any particular TOC. Perhaps one person runs a tool to build the topics, another person identifies which topics to include, and then someone else runs a tool to build a TOC from identified topics.

No matter how you create the TOC, the basic pattern is to identify something to include in the TOC, determine the relative placement of that thing in the hierarchy, and insert markup into the XML Flare TOC file to reflect those choices.

To create a TOC you need to manipulate XML. The Flare interface adds some nice visual conventions to help you do this. For example, you can drag a topic from the Content window into a TOC. In fact, you can select multiple topics and drag them into a TOC as a group. Flare also enables you to auto-generate a TOC from a set of topics in a project.

You can replicate these features outside the user interface, but that isn't very interesting or useful. Adding functionality is more interesting. How about an algorithm to sort a Flare TOC? Or what about program to insert filler topics in between all of the top-level nodes in a TOC?

Why would you want to insert filler topics? Imagine you have created a sequence of topics you normally want to be viewed in a particular order. However, you also want those topics to stand alone so they can be understood without outside context. You probably need transitional content between each topic in the sequence, but you probably also want to avoid inserting that transitional content at the beginning of each topic. One way to overcome the obstacle imposed by a strict topic-based approach is to use filler topics to provide the transitional context.

Imagine a tool that generates filler topics for a browse sequence in a training manual. That is, text like, "You've reviewed Pre-configuring. Now learn about Starting the Machine." And while you are at it, why not insert quote blocks in those filler topics with the text of the first paragraphs from the previous and next topics? Maybe you will go a step beyond that and turn each of those paragraphs into snippets and refer to those snippets from the original topics and the filler topics. All of this can be done programmatically.

Imagine a process that takes data from a database, a spreadsheet, or perhaps a web service and creates Flare TOCs based on that data. If you generate Flare topics, the cost of automating the process of adding those topics to a Flare TOC is low, not many lines of code at all. Then you not only have topics in a folder, but you also have an additional organizing artifact, the TOC, with which to work.

Processes such as these can give you big savings because you can eliminate manual work. A program can help you and your team avoid mindless hours spent cutting and pasting text, creating new files, typing titles, and inserting files into TOCs. Of course, not everything can be automated and automation is not free, but if you weigh the costs of a manual solution against an automated solution, you can make an informed decision.

TOCs in output

Export behavior
Flare TOCs usually become a TOC or Contents navigation in web-based outputs. Typically this is a hierarchical navigation panel to the left of the screen, as is common with help systems. However, skins let you define other configurations, and they can even hide the TOC completely.

In print-based output, there is usually some visual indication of the TOC structure. The heading level in the output is based on the position of either the topics in the Flare TOC or the heading elements in the topic files. This selection is made in the target file. TOCs may also appear in PDF bookmarks or Word document maps.

For either web or print output, if you use proxies such as the TOC proxy or mini-TOC proxy, TOC hierarchy information appears inline with other content. For example, if you place a TOC proxy at the beginning of the Flare TOC, Flare will place a table of contents at that location when you generate output.

HTML5 and web-based output
In web-based outputs, TOCs are a means of navigation within the information system. These outputs typically include a left-hand table of contents with some variation of expanding and collapsing entries. Depending on the configuration you use, you can place the TOC somewhere else on the page or hide it entirely.

Other end-user navigation tools, such as mini-TOCs and breadcrumbs, use the TOC you specify in the target. The main difference between proxy-based navigation and the navigation pane is that the navigation pane is built dynamically from data files and the proxies are transformed into generated links in topics.

For example, prior to version 11, MadCap Software's online help for Flare was HTML5 generated from a Flare target, with navigation dynamically rendered from *JSON*. However, other navigation aids within the topic pane were static links generated in place of a proxy.[2]

In Flare's WebHelp output, the TOC is contained in XML data files. This was also the case for HTML5 output prior to version 9 of Flare. However, with version 9, the data file was switched to a JavaScript file containing `define()` statements that encapsulate the JSON TOC data. The `define()` function is part of the RequireJS JavaScript library.

Flare includes the RequireJS library in the script tags in the header of HTML5 output. Example 5.6 shows the script tags in the header of an HTML5 output main page (as of version 9).

Example 5.6 – Script tags in an HTML5 main page (as of version 9)

```
<script type="text/javascript" src="Resources/Scripts/jquery.min.js">
</script>
<script type="text/javascript" src="Resources/Scripts/plugins.min.js">
</script>
<script type="text/javascript" src="Resources/Scripts/require.min.js">
</script>
<script type="text/javascript" src="Resources/Scripts/require.config.js">
</script>
<script type="text/javascript" src="Resources/Scripts/MadCapAll.js">
</script>
```

To read the data files in WebHelp or HTML5 you need access to the XML or JavaScript and an algorithm to interpret the data. We explore that next.

Exploring the HTML5 TOC data files

The TOC data in HTML5 output is chunked; there is one main `Toc.js` file, which refers to the other chunk files. This means you must first read the `Toc.js` file and then access data from the

---

[2] The current version of MadCap online help (http://webhelp.madcapsoftware.com) still uses HTML5 and JSON, but no longer uses the tri-pane help model.

other chunk files as necessary. Whether your application reads all of the chunk files at once or only as needed is a design decision, which is usually based on memory use and load times.

Example 5.7 – `Toc.js` file

```
define({
    numchunks:1,prefix:'Toc_Chunk',
    chunkstart:['/Content/Topic 7.htm'],
    tree:{n:[{i:0,c:0,n:[{i:1,c:0}]},
    {i:2,c:0,n:[{i:3,c:0,n:[{i:4,c:0}]},{i:5,c:0},{i:6,c:0}]},{i:7,c:0}]}
});
```

Example 5.7 shows the contents of a `Toc.js` file generated for an HTML5 TOC. The `define()` function loads the JSON data for the TOC.

With a little knowledge of RequireJS and developer tools in a web browser, you can discover what modules have been loaded. To see what we mean, open an HTML5 output in Google Chrome,[3] open the console (**View → Developer → JavaScript Console**), type the following, and press Enter.

`require.s.contexts._.defined`

Something like this should appear:

`Object {Data/Toc.js: Object, Data/Toc_Chunk0.js: Object}`

This is expandable information about the objects for the RequireJS modules for the TOC described in `Toc.js` and the chunk file `TocChunk0.js`, which have been loaded by the main page JavaScript. You may find it useful to explore how the skin takes it from there. But if you want to use just the data files for the TOC in an HTML5 output, you can skip all that. Using the data files in a generated output doesn't create a dependency on anything else in the output.

If your page is going to access these files, the easiest route is to use RequireJS. You can use the copy in the output, get another copy, or link to a hosted location. The relative path for those files in an HTML5 output is given in the `<script>` elements for the main help page in the output (see Example 5.8).

---

[3] Any HTML5 output generated by Flare will work or you can open http://webhelp.madcapsoftware.com/flare11

Example 5.8 – HTML5 script tags for `require.min.js` **and** `require.config.js`

```
<script type="text/javascript" src="Resources/Scripts/require.min.js">
</script>
<script type="text/javascript" src="Resources/Scripts/require.config.js">
</script>
```

If you want to use the `Toc.js` module, you can use the `require()` function as shown here:

```
require(['../Data/Toc.js'], function(toc) {});
```

You can define an action inside the curly brackets that follow the anonymous function, `function(toc)`. Any code you place there can access what is defined in `Toc.js`. For example, you can display the number of TOC chunks in a popup, as shown here:

```
require(['../Data/Toc.js'], function(toc) {alert(toc.numchunks);});
```

Looking at Example 5.7, the first line (`numchunks:1,prefix:'Toc_Chunk'`) tells us that the chunk file names have the prefix string `Toc_Chunk` followed by a zero-based index. Thus, the first file name is `Toc_Chunk0.js`. If numchunks was 2, then there would be two files: `Toc_Chunk0.js` and `Toc_Chunk1.js`.

If you want to read each chunk in order, all you have to do is iterate from 0 to `numchunks-1` and read the file name described by `<prefix><iteration>.js`.

You could also just name each file in the `require()` function. For example, for a single chunk, `require()` could look like this.

```
require(['../Data/Toc.js', '../Data/Toc_Chunk0.js'], function(toc, chunk) {});
```

But if you do that, you then have to hardcode the name of the chunk file for every case.

In Example 5.7, beginning with the second line in the `define()` function, the code specifies a topic (`chunkstart`) and a tree. Example 5.9 shows how you can traverse the tree in `Toc.js`. Adjust the path to `Toc.js` as necessary.

Example 5.9 – Code to traverse the tree in `Toc.js`

```
require(['../Data/Toc.js'], function(toc) {
    function traverseNodes(n, topNode) {
        $.each(n, function(key, value) {
            if (typeof value.n !== "undefined") {
                traverseNodes(value.n);
            }
        });
    }
    traverseNodes(toc.tree.n);
});
```

Let's look at the JSON TOC in more detail using the following snippet from Example 5.7:

`n:[{i:0,c:0,n:`

The snippet begins with `n:[`, which represents the root of the tree. Then there are two name/value pairs that identify the first node. The first name/value pair (`i:0`) represents an index `i` with the value `0` (the indexes are all zero based). The second name/value pair (`c:0`) refers to the chunk file (in this case `Toc_Chunk0.js`) that contains the node. There is an optional name/value pair with the name `n` that nests descriptions for child nodes.

This first node, then, refers to the node with `i` equal to `0` in the file `Toc_Chunk0.js`. The name/value pair for this node in `Toc_Chunk0.js` is the following:

`'/Content/Topic1.htm':{i:[0],t:['Topic 1'],b:['']}`

The name of this node (`'/Content/Topic1.htm'`) is the relative path to the file that contains the topic. The value contains three name/value pairs: `i:[0]` is the index, `t:['Topic 1']` contains the topic label, and `b:['']` contains an optional bookmark. To access a topic when traversing `Toc.js`, you need to locate the applicable chunk file using the name/value pair `c:X` and the node within that file that matches the index identified in the name/value pair `i:Y`.

Print output

In print output the conventions are similar to web output. For example, PDF has TOC-based navigation built in. When viewed in Adobe Acrobat, the TOC displays as a set of bookmarks that, when clicked, move the page view to the applicable bookmark. As with web-based outputs, proxies appear in line with the content. For example, a topic containing a TOC proxy can be used to create a table of contents at the beginning of a document.

TOC code samples: creating and modifying a TOC

XSLT example

You may find it useful to append a string to the label for every TOC entry. Example 5.10 is an XSLT example which adds "_topic" to the end of the label in each TOC entry in a TOC. You could also create one to undo that action. You might want to do something like this if your workflow involves releasing beta content. Just replace "_topic" with "_beta" in the example.

Example 5.10 – XSLT that appends "_topic" to each entry in a TOC

```
<?xml version="1.0" encoding="utf-8"?>
<xsl:stylesheet version="1.0"
                xmlns:xsl="http://www.w3.org/1999/XSL/Transform"
                xmlns:msxsl="urn:schemas-microsoft-com:xslt"
                exclude-result-prefixes="msxsl">
  <xsl:output method="xml" indent="yes"/>

  <xsl:template match="@*|node()">
    <xsl:copy>
      <xsl:apply-templates select="@*|node()"/>
    </xsl:copy>
  </xsl:template>
  <xsl:template match="@Title">
    <xsl:attribute name="Title">
      <xsl:value-of select="concat(.,'_topic')"/>
    </xsl:attribute>
  </xsl:template>
</xsl:stylesheet>
```

.NET examples

Sometimes authors build an outline before writing the content for each section in the outline. Example 5.11 is a .NET example that creates a Flare TOC based on a list of topics in a CSV file. The first field on each line in the CSV file contains the title and the second (last) field contains text. The result is a TOC with an entry for each row in the CSV file. Each entry links to a topic that contains two snippets. The first snippet contains text from the second field, and the second snippet contains `<p>Default text</p>`. This example is described in more detail in a blog entry at tregner.com.[4] Note that this example has hard-coded values that should be passed in as arguments, and it does not check for the existence of files and directories.

---

[4] http://tregner.com/flare-blog/combining-generated-and-flare-managed-content

Example 5.11 – Build TOC from CSV list

```
Module Module1

  Public Sub CreateSnippet(ByVal fileName As String, ByVal description As String)
    Dim Snippet As XDocument = _
      <?xml version="1.0" encoding="utf-8"?>
      <html xmlns:MadCap="http://www.madcapsoftware.com/Schemas/MadCap.xsd">
        <head>
        </head>
        <body>
          <p><%= description %></p>
        </body>
      </html>
    Snippet.Save("C:\Example\Content\Resources\Snippets\" & fileName)
  End Sub

  Public Sub CreateTopic(ByVal fileName As String, ByVal topicName As String)
    Dim Topic As XDocument = _
      <?xml version="1.0" encoding="utf-8"?>
      <html xmlns:MadCap="http://www.madcapsoftware.com/Schemas/MadCap.xsd">
        <head><title><%= topicName %></title>
        </head>
        <body>
          <h1><%= topicName %></h1>
          <MadCap:snippetBlock src=<%= "Resources/Snippets/" _
                        & topicName & "-Refreshable.flsnp" %>/>
          <MadCap:snippetBlock src=<%= "Resources/Snippets/"_
      & topicName & "-EditInFlare.flsnp" %>/>
        </body>
      </html>
    Topic.Save("C:\Example\Content\" & fileName)
  End Sub

  Sub Main()
    Dim FlareToc As XDocument = _
      <?xml version="1.0" encoding="utf-8"?>
      <CatapultToc></CatapultToc>

    Using MyReader As New FileIO.TextFieldParser("C:\Example\wonderful-things.csv")
      MyReader.TextFieldType = FileIO.FieldType.Delimited
      MyReader.SetDelimiters(",")
      Dim currentRow As String()
      While Not MyReader.EndOfData
        currentRow = MyReader.ReadFields()
        CreateTopic(currentRow.First & ".htm", currentRow.First)
        CreateSnippet(currentRow.First & "-Refreshable.flsnp", currentRow.Last)
        CreateSnippet(currentRow.First & "-EditInFlare.flsnp", "Default text")
        FlareToc.Add()
        FlareToc.Root.Add(<TocEntry Link=<%= "/Content/" _
```

```
                    & currentRow.First _
                    & ".htm" %> Title=<%= currentRow.First %>>
            </TocEntry>)
    End While
    End Using
    FlareToc.Save("C:\Example\Project\TOCs\wonderful-things.fltoc")
  End Sub
End Module
```

Another useful feature would be the ability to sort TOCs. Suppose you have a complex set of topics with a TOC ordered so that each topic builds on the previous topic. You might want to offer a secondary TOC that is ordered alphabetically. Figure 5.1 shows a .NET Windows Forms application that lets the reader sort a Flare TOC in various ways. Example 5.12 shows the matching Visual Basic code.

Figure 5.1 – UI for a custom TOC sorter

Flare Tables of Contents

Example 5.12 – Visual Basic logic for custom TOC sorter

```vb
Imports System.Xml

Public Class Form1

  Private NewToc As XDocument

    Private Sub ButtonOpen_Click(sender As System.Object,
                e As System.EventArgs) Handles ButtonOpen.Click
        OpenFileDialogToc.ShowDialog()
    End Sub

    Private Sub ButtonSave_Click(sender As System.Object,
                e As System.EventArgs) Handles ButtonSave.Click
        If String.IsNullOrWhiteSpace(OpenFileDialogToc.FileName) Or
 Not My.Computer.FileSystem.FileExists(OpenFileDialogToc.FileName) Then
            MsgBox("A file has not been selected.")
        Else
            SaveFileDialogToc.ShowDialog()
        End If
    End Sub

    Private Sub OpenFileDialogToc_FileOk(sender As System.Object,
                e As System.ComponentModel.CancelEventArgs) _
        Handles OpenFileDialogToc.FileOk
        Dim Toc As XmlDocument = New XmlDocument()
        Toc.Load(OpenFileDialogToc.FileName)
        RefreshTree(Toc)
        NewToc = XDocument.Load(OpenFileDialogToc.FileName)
    End Sub

  Private Sub RefreshTree(ByVal Toc As XmlDocument)
    TreeViewToc.Nodes.Clear()
    TreeViewToc.Nodes.Add(New TreeNode())
    Dim TocNode As TreeNode = New TreeNode()
    TocNode = TreeViewToc.Nodes(0)
    AddNodeToDisplay(Toc.DocumentElement, TocNode)
    TreeViewToc.ExpandAll()
  End Sub

  Private Sub AddNodeToDisplay(ByVal InTocFileNode As XmlNode,
                               ByVal InTocTreeNode As TreeNode)
    Dim TocFileNode As XmlNode
    Dim TocTreeNode As TreeNode
    Dim TocFileNodeList As XmlNodeList

    If InTocFileNode.HasChildNodes Then
      TocFileNodeList = InTocFileNode.ChildNodes
      For i As Integer = 0 To TocFileNodeList.Count - 1 Step 1
```

```
          TocFileNode = InTocFileNode.ChildNodes(i)
          InTocTreeNode.Nodes.Add(_
            New TreeNode(TocFileNode.Attributes("Title").Value))
          TocTreeNode = InTocTreeNode.Nodes(i)
          AddNodeToDisplay(TocFileNode, TocTreeNode)
        Next
      Else
        InTocTreeNode.Text = InTocFileNode.Attributes("Title").Value
      End If
    End Sub

    Private Sub Flatten(ByVal Toc As String)
        If String.IsNullOrWhiteSpace(OpenFileDialogToc.FileName) Or
 Not My.Computer.FileSystem.FileExists(OpenFileDialogToc.FileName) Then
            MsgBox("A file has not been selected.")
        Else
            Dim TocXml As XDocument = XDocument.Load(Toc)
            Dim FlatToc As XDocument = _
                <?xml version="1.0" encoding="utf-8"?>
                <CatapultToc
                Version="1">
        </CatapultToc>

            For Each element In TocXml.Root.Descendants
                Dim element2 As XElement = _
                    <TocEntry></TocEntry>
                For Each attr In element.Attributes
                    element2.SetAttributeValue(attr.Name, attr.Value)
                Next
                FlatToc.Root.Add(element2)
            Next

            NewToc = FlatToc
            LabelChangeApplied.Text = "Flattened"

            Dim xd As New XmlDocument
            Dim xr = FlatToc.CreateReader()
            xd.Load(xr)
            RefreshTree(xd)
        End If
    End Sub

    Private Sub SortTopNodes(ByVal Toc As String)
        If String.IsNullOrWhiteSpace(OpenFileDialogToc.FileName) Or
 Not My.Computer.FileSystem.FileExists(OpenFileDialogToc.FileName) Then
            MsgBox("A file has not been selected.")
        Else
            Dim TocXml As XDocument = XDocument.Load(Toc)
            Dim SortedToc As XDocument = _
                <?xml version="1.0" encoding="utf-8"?>
```

```vb
                    <CatapultToc
                        Version="1">
                </CatapultToc>

Dim ChildrenOfCatapultToc = From xElement In TocXml.Root.Elements _
   Order By CStr(xElement.Attribute("Title")) Ascending _
   Select xElement

                SortedToc.Root.Add(ChildrenOfCatapultToc)

                NewToc = SortedToc
                LabelChangeApplied.Text = "Top nodes sorted"

                Dim xd As New XmlDocument
                Dim xr = SortedToc.CreateReader()
                xd.Load(xr)
                RefreshTree(xd)
        End If
    End Sub
    Private Sub SortInnerNodes(ByVal Toc As String)

      If String.IsNullOrWhiteSpace(OpenFileDialogToc.FileName) Or
Not My.Computer.FileSystem.FileExists(OpenFileDialogToc.FileName) Then
        MsgBox("A file has not been selected.")
      Else
        Dim TocXml As XDocument = XDocument.Load(Toc)
        Dim SortedToc As XDocument = _
          <?xml version="1.0" encoding="utf-8"?>
          <CatapultToc
            Version="1">
          </CatapultToc>

        SortTocEntry(TocXml.Root)

        SortedToc.Root.ReplaceWith(TocXml.Root)

        NewToc = SortedToc
        LabelChangeApplied.Text = "All nodes sorted"

        Dim xd As New XmlDocument
        Dim xr = SortedToc.CreateReader()
        xd.Load(xr)
        RefreshTree(xd)
      End If
    End Sub

    Private Sub SortTocEntry(ByVal tocEntry As XElement)
      If tocEntry.HasElements Then
        For Each entry In tocEntry.Elements
          SortTocEntry(entry)
```

```
    Next
    tocEntry.ReplaceNodes(From xElement In tocEntry.Elements _
        Order By CStr(xElement.Attribute("Title")) Ascending _
        Select xElement)
    End If
End Sub

    Private Sub ButtonFlatten_Click(sender As System.Object,
              e As System.EventArgs) _
        Handles ButtonFlatten.Click
        Flatten(OpenFileDialogToc.FileName)
    End Sub

    Private Sub SaveFileDialogToc_FileOk(sender As System.Object,
              e As System.ComponentModel.CancelEventArgs) _
        Handles SaveFileDialogToc.FileOk
        NewToc.Save(SaveFileDialogToc.FileName)
    End Sub

    Private Sub ButtonSortTopNodes_Click(sender As System.Object,
              e As System.EventArgs) _
        Handles ButtonSortTopNodes.Click
        SortTopNodes(OpenFileDialogToc.FileName)
    End Sub

    Private Sub ButtonSortInnerNodes_Click(sender As System.Object,
              e As System.EventArgs) _
        Handles ButtonSortInnerNodes.Click
        SortInnerNodes(OpenFileDialogToc.FileName)
    End Sub

    Private Sub Form1_Load(sender As Object,
              e As EventArgs) Handles MyBase.Load

    End Sub
End Class
```

JavaScript example

You can also manipulate a TOC in web-based outputs using JavaScript. Example 5.13 is a JavaScript example that creates a button to sort the top level nodes in an HTML5 TOC. This example uses the same technique used in Example 5.12, but applied to an output TOC, which uses JSON, rather than the source TOC, which uses XML.

Example 5.13 – Top level HTML5 output TOC sort with JavaScript

```
var topNodes = document.getElementsByClassName("tree")[0].childNodes;
var nodeHtmlAndClasses = [];

for (var i = 0, l = topNodes.length; i < l; i++) {
    var nodeHtmlAndClass = { iH: topNodes[i].innerHTML,
  cN: topNodes[i].className, lbl: topNodes[i].textContent };
    nodeHtmlAndClasses.push(nodeHtmlAndClass);
}
nodeHtmlAndClasses.sort(function (a, b) {
    var labelA = a.lbl.toLowerCase(), labelB = b.lbl.toLowerCase()
    if (labelA < labelB)
        return -1
    if (labelA > labelB)
        return 1
    return 0
});
for (var i = 0, l = topNodes.length; i < l; i++) {
    topNodes[i].innerHTML = nodeHtmlAndClasses[i]["iH"];
    topNodes[i].className = nodeHtmlAndClasses[i]["cN"];
}
```

Java example

Example 5.14 shows a Java application that creates Flare topics and TOCs. The methods `returnTopic()`, `returnEmptyToc()`, and `returnTocEntry()` create XML for topics, TOCs, and `<TocEntry>` elements, respectively. The methods `returnTocEntryAppendChildEntry()`, `returnFlatToc()`, and `returnTocEntryAppendTocEntry()` connect these artifacts together. The logic in the main method uses these methods and additional XML manipulation to create topics, add content to them, and organize them in a TOC.

Example 5.14 – An approach to creating Flare topics and TOCs with Java

```
package flarecontentgenerator;

import java.io.File;
import java.io.IOException;
import java.io.StringReader;
import java.util.List;

import javax.xml.parsers.DocumentBuilder;
import javax.xml.parsers.DocumentBuilderFactory;
import javax.xml.parsers.ParserConfigurationException;
```

```java
import javax.xml.transform.Transformer;
import javax.xml.transform.TransformerException;
import javax.xml.transform.TransformerFactory;
import javax.xml.transform.dom.DOMSource;
import javax.xml.transform.stream.StreamResult;

import org.w3c.dom.Document;
import org.w3c.dom.Element;
import org.xml.sax.InputSource;
import org.xml.sax.SAXException;

public class FlareContentGenerator {

 public static Document returnTopic() throws ParserConfigurationException,
   SAXException, IOException {
  DocumentBuilderFactory documentFactory = DocumentBuilderFactory
    .newInstance();
  DocumentBuilder documentBuilder = documentFactory.newDocumentBuilder();

  String topicString = "<?xml version=\"1.0\" encoding=\"utf-8\"?>"
    + "<html xmlns:MadCap=\"http://www.madcapsoftware.com/Schemas/MadCap.xsd\">"
    + "<head />" + "<body />" + "</html>";

  Document topic = documentBuilder.parse(new InputSource(
    new StringReader(topicString)));
  return topic;
 }

 public static Document returnTopic(String title, String text)
   throws ParserConfigurationException, SAXException, IOException {
  Document topic = returnTopic();
  Element titleElement = topic.createElement("title");
  titleElement.setTextContent(title);
  topic.getElementsByTagName("head").item(0).appendChild(titleElement);
  Element headingElement = topic.createElement("h1");
  headingElement.setTextContent(title);
  topic.getElementsByTagName("body").item(0).appendChild(headingElement);
  Element paragraph = topic.createElement("p");
  paragraph.setTextContent(text);
  topic.getElementsByTagName("body").item(0).appendChild(paragraph);
  return topic;
 }

 public static Document returnEmptyToc()
   throws ParserConfigurationException, SAXException, IOException {
  DocumentBuilderFactory documentFactory = DocumentBuilderFactory
    .newInstance();
  DocumentBuilder documentBuilder = documentFactory.newDocumentBuilder();

  String tocString = "<?xml version=\"1.0\" encoding=\"utf-8\"?>"
```

```
      + "<CatapultToc Version=\"1\" />";
  Document toc = documentBuilder.parse(new InputSource(new StringReader(
    tocString)));
  return toc;
}

public static Document returnFlatToc(List<Element> tocEntries)
    throws ParserConfigurationException, SAXException, IOException {
  Document toc = returnEmptyToc();
  for (Element t : tocEntries) {
    toc.getElementsByTagName("CatapultToc").item(0).appendChild(t);
  }
  return toc;
}

public static Element returnTocEntry(Document toc, String topicLabel,
    String topicLink) {
  Element tocEntry = toc.createElement("TocEntry");
  tocEntry.setAttribute("Title", topicLabel);
  tocEntry.setAttribute("Link", topicLink);
  return tocEntry;
}

public static Element returnTocEntryAppendChildEntry(Document toc,
    Element tocEntry, String topicLabel, String topicLink) {
  Element childTocEntry = returnTocEntry(toc, topicLabel, topicLink);
  tocEntry.appendChild(childTocEntry);
  return tocEntry;
}

public static Element returnTocEntryAppendChildEntry(Document toc,
    Element tocEntry, Element childTocEntry) {
  tocEntry.appendChild(childTocEntry);
  return tocEntry;
}

public static Document returnTocAppendTocEntry(Document toc, String topicLabel,
    String topicLink) {
  Element tocEntry = returnTocEntry(toc, topicLabel, topicLink);
  toc.getElementsByTagName("CatapultToc").item(0).appendChild(tocEntry);
  return toc;
}

public static Document returnTocAppendTocEntry(Document toc,
    Element tocEntry) {
  toc.getElementsByTagName("CatapultToc").item(0).appendChild(tocEntry);
  return toc;
}
```

```java
public static void main(String[] args) throws ParserConfigurationException,
   SAXException, IOException, TransformerException {
 Document firstTopic = returnTopic();
 String firstTopicTitle = "First Topic";
 String firstTopicFilename = "FirstTopic.htm";
 String firstTopicLink = "/Content/" + firstTopicFilename;
 String firstTopicText = "This is some text for the first topic.";
 firstTopic = returnTopic(firstTopicTitle, firstTopicText);

 String secondTopicTitle = "Second Topic";
 String secondTopicFilename = "SecondTopic.htm";
 String secondTopicLink = "/Content/" + secondTopicFilename;
 String secondTopicText = "This is some text for the second topic.";
 Document secondTopic = returnTopic(secondTopicTitle, secondTopicText);

 Document toc = returnEmptyToc();
 Element firstTocEntry = returnTocEntry(toc, firstTopicTitle,
   firstTopicLink);
 Element secondTocEntry = returnTocEntry(toc, secondTopicTitle,
   secondTopicLink);
 Element thirdTocEntry = returnTocEntryAppendChildEntry(toc,
   firstTocEntry, secondTocEntry);
 toc = returnTocAppendTocEntry(toc, thirdTocEntry);

 String basePathContent = "C:\\Examples\\ExampleProject\\Content\\";
 String basePathTOCs = "C:\\Examples\\ExampleProject\\Project\\TOCs\\";
 TransformerFactory transformerFactory = TransformerFactory
   .newInstance();
 Transformer transformer = transformerFactory.newTransformer();
 DOMSource source = new DOMSource(firstTopic);
 File file = new File(basePathContent + firstTopicFilename);
 StreamResult streamResult = new StreamResult(file);
 transformer.transform(source, streamResult);

 source = new DOMSource(secondTopic);
 file = new File(basePathContent + secondTopicFilename);
 streamResult = new StreamResult(file);
 transformer.transform(source, streamResult);

 Document emptyToc = returnEmptyToc();
 source = new DOMSource(emptyToc);
 streamResult = new StreamResult(new File(basePathTOCs + "empty.fltoc"));
 transformer.transform(source, streamResult);

 source = new DOMSource(toc);
 streamResult = new StreamResult(new File(basePathTOCs + "other.fltoc"));
 transformer.transform(source, streamResult);
 }
}
```

CHAPTER 6
Flare Indexes, Glossaries, and Search

Flare supports indexes, glossaries, and search. Indexes and glossaries can be rendered in different ways. For example, glossary terms can be displayed as footnotes in print output or as pop ups in web-based output. Glossaries and indexes can be included as tabbed content in web-based output. And for any output type, you can use index and glossary proxies to place an index or a glossary at a certain location in your content. Search is a separate functionality for web-based outputs. Search behavior varies depending on the particular output type.

Flare indexes

A Flare index is a collection of links to markers spread throughout a set of content. You can place an index proxy at the point in your content where you want the index to appear, or with web-based outputs, you can display an index as part of the skin's navigation. Figure 6.1 shows an index in printed output and Figure 6.2 shows one in a web-based skin.

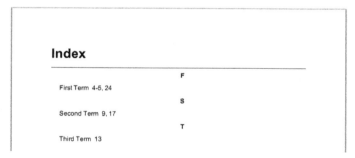

Figure 6.1 – Index in a PDF

Figure 6.2 – Index in HTML5 tri-pane output

Index proxies

Index proxies are MadCap elements that you place in your content to define where an index should be rendered when you build a target. For example, if a topic that contains an index proxy is placed in a TOC that is organized as chapters in a PDF or XPS output, the index will appear in the chapter that contains that topic. If you include the same topic in a web-based output, the index will appear in the topic file that contains the index proxy.

Like other proxies, an index proxy appears as a grey block in the XML Editor. The markup for an index proxy looks like this:

```
<MadCap:indexProxy style="mc-index-headings: true;" />
```

Index link sets

Index link sets enable you to define See also, Sort as, and See behaviors in generated indexes.

Example 6.1 – XML for a `<CatapultIndexLinkSet>` element with no entries

```
<?xml version="1.0" encoding="utf-8"?>
<CatapultIndexLinkSet></CatapultIndexLinkSet>
```

Auto Index Sets

Flare can generate an index for you when it builds the target. When you use Auto Index Sets, you define a list of terms and the build process finds those terms and creates the index automatically.

However, it is useful to know how to programmatically find and insert index markers. For example, you can extend a process that traverses a TOC and modifies the topics to also search for terms. The next section describes how to create an Auto Index Set from another data source.

Create an Auto Index Set

This section describes a .NET utility that reads topics in a given folder, identifies all the words in those topics, sorts them by their frequency of occurrence, and then creates an Auto Index Set that contains those words. Example 6.2 and Example 6.3 implement this capability using Visual Basic.

This utility can be used as the starting point for building an effective Auto Index Set. Using the generated Auto Index Set as a starting point, you can selectively remove entries you don't need. You could also create a more powerful version of the utility that excludes terms that don't appear in a dictionary or that do appear in a list of terms you don't want in the index.

Example 6.2 – WordDetail class

```
Public Class WordDetail
    Property Word As String
    Property Frequency As Integer

    Sub New(NewWord As String, NewFrequency As Integer)
        Word = NewWord
        Frequency = NewFrequency
    End Sub
End Class
```

Example 6.3 – Module to create an Auto Index Set

```
Imports System
Imports System.IO
Imports System.Xml.Linq
Imports System.Text.RegularExpressions

Module Module1

Private Function HasWord(Word As String,
                 WordDetailList As List(Of WordDetail)) As Boolean
  For Each wd In WordDetailList
    If wd.Word = Word Then
      Return True
    End If
  Next
  Return False
End Function

Private Sub AddWordDetail(Word As String,
        ByRef WordDetailList As List(Of WordDetail))
  WordDetailList.Add(New WordDetail(Word, 1))
End Sub

Private Sub UpdateWordDetail(Word As String,
        ByRef WordDetailList As List(Of WordDetail))
  For Each wd In WordDetailList
    If wd.Word = Word Then
      wd.Frequency = wd.Frequency + 1
      Exit For
    End If
  Next
End Sub
```

Flare Indexes, Glossaries, and Search

```vb
Private Sub SortWordDetailListByFrequency(ByRef WordDetailList As _
                                List(Of WordDetail))
  WordDetailList = WordDetailList.OrderBy(Function(x) x.Frequency).ToList()
End Sub

Sub Main(ByVal args() As String)
  Dim ContentFolder = args(0)
  Dim AutoIndexSetFilePath = args(1)
  Dim WordDetailList As List(Of WordDetail) = New List(Of WordDetail)
  Dim DirInfo As New DirectoryInfo(ContentFolder)
  Dim FilesArray As FileInfo() = DirInfo.GetFiles()
  For Each f In FilesArray
    If f.Extension.Equals(".htm") Then
      Dim Topic As XDocument = XDocument.Load(f.FullName)
      Dim TopicTextString As String = Topic.Root.Element("body").Value
      TopicTextString = Regex.Replace(TopicTextString, "[^\w\.@-]", " ")
      Dim TopicText = TopicTextString.ToCharArray
      Dim NotDone As Boolean = True
      Dim StartIndex As Integer = 0
      Dim CurrentIndex = 0
      Dim EndIndex As Integer = 0
      If TopicText.Length = 0 Then
        NotDone = False
      End If
      While NotDone And CurrentIndex < TopicText.Length
        If Char.IsWhiteSpace(TopicText(CurrentIndex)) Then
          StartIndex = CurrentIndex
          CurrentIndex = CurrentIndex + 1
        Else
          Dim NotFound As Boolean = True
          While NotDone And CurrentIndex < TopicText.Length And _
              NotFound
            If Char.IsWhiteSpace(TopicText(CurrentIndex)) Or _
                         CurrentIndex = TopicText.Length Then
              If CurrentIndex = TopicText.Length Then
                NotDone = False
              End If
              EndIndex = CurrentIndex
              Dim Word As String = ""
              Dim WordArray(EndIndex - StartIndex) As Char
              Array.Copy(TopicText, StartIndex, _
                         WordArray, 0, EndIndex - StartIndex)
              Word = New String(WordArray)
              Word = Regex.Replace(Word, "[^\w\.@-]", "")
              If Word.LastIndexOf(".") = Word.Length - 1 Or _
                    Word.LastIndexOf(",") = Word.Length - 1 Then
                Word = Word.Substring(0, Word.Length - 1)
              End If
              If HasWord(Word, WordDetailList) Then
                UpdateWordDetail(Word, WordDetailList)
```

```
            Else
              AddWordDetail(Word, WordDetailList)
            End If
            NotFound = False
          Else
            CurrentIndex = CurrentIndex + 1
          End If
        End While
      End If
    End While
  End If
Next f

Dim AutoIndexSet As XDocument = _
   <?xml version="1.0" encoding="utf-8"?>
   <CatapultAutoIndexSet></CatapultAutoIndexSet>

SortWordDetailListByFrequency(WordDetailList)

For Each wd In WordDetailList
  Try
    If wd.Word IsNot Nothing Then
      AutoIndexSet.Root.Add(
        <AutoIndex Phrase=<%= wd.Word.ToString %>
        IndexTerm=<%= wd.Word.ToString %>/>)
    End If
  Catch ex As Exception

  End Try
Next

  AutoIndexSet.Save(AutoIndexSetFilePath)
End Sub
End Module
```

Flare glossaries

Glossaries are similar to indexes in how they are managed and how they appear in various outputs. Glossaries are key/value pairs that relate a term to a topic or a short definition of the term. Like indexes, you can place a glossary inline with your content via a proxy, or with web-based outputs, you can select to display a glossary as part of the skin's navigation.

Glossary proxies

Like index proxies, glossary proxies are MadCap elements that you place in your content to define where a glossary should be rendered when you build a target. For example, if a topic that contains a glossary proxy is placed in a TOC that is organized as chapters in a PDF or XPS output, the glossary will appear in the chapter that contains the topic. If you include the same topic in a web-based output, the glossary will appear in the topic file that contains the index proxy.

Glossary files

Glossary files are where you define the key/value pairs that Flare uses to create a glossary when a target is built.

Example 6.4 – XML markup for a glossary file

```
<?xml version="1.0" encoding="utf-8"?>
<CatapultGlossary>
 <GlossaryEntry>
  <Terms>
   <Term>My Term</Term>
  </Terms>
  <Definition>My definition</Definition>
 </GlossaryEntry>
</CatapultGlossary>
```

Creating and modifying glossaries programmatically

Building on Example 6.3, where we created an Auto Index Set with a .NET utility, we could modify the code to create a skeletal glossary of terms ordered by frequency. It is harder to programmatically extract a definition from the content (that is an artificial intelligence problem that is a little outside the scope of this book). However, a skeletal list of terms is a better starting point than nothing.

Example 6.5 adds a term to the first glossary entry in a glossary.

Example 6.5 – .NET code to add a term to the glossary

```
using System;
using System.Collections.Generic;
using System.Linq;
using System.Text;
using System.Threading.Tasks;
using System.Xml.Linq;

namespace AlterGlossary
{
    class AlterGlossary
    {
        static void Main(string[] args)
        {
            //args[0] is the path to the glossary
            //args[1] is the term to add
            XDocument glossary = XDocument.Load(args[0]);
            XElement term = new XElement("Term");
            term.Value = args[1];
            glossary.Descendants("Terms").First().Add(term);
            glossary.Save(args[0]);
        }
    }
}
```

Web output JavaScript example

And to round out the examples, perhaps you want to sort the glossary entries in an HTML5 tri-pane output differently than they are sorted by default. Example 6.6 sorts the glossary entries in reverse alphabetical order. This script can be run on a generated HTML5 output. It could be attached to a button or some other mechanism. But if you just want to see it work, you can run the script directly from your browser's developer console.

Example 6.6 – JavaScript to sort an HTML5 glossary in reverse alphabetical order

```
var glossary = document.getElementById("glossary");
var topNodes = glossary.getElementsByClassName("tree")[0].childNodes;
var nodeHtmlAndClasses = [];

for (var i = 0, l = topNodes.length; i < l; i++) {
    var nodeHtmlAndClass = { iH: topNodes[i].innerHTML,
  cN: topNodes[i].className, lbl: topNodes[i].textContent };
    nodeHtmlAndClasses.push(nodeHtmlAndClass);
}

nodeHtmlAndClasses.sort(function (a, b) {
    var labelA = a.lbl.toLowerCase(), labelB = b.lbl.toLowerCase()
    if (labelA > labelB)
        return -1
    if (labelA < labelB)
        return 1
    return 0
});

for (var i = 0, l = topNodes.length; i < l; i++) {
    topNodes[i].innerHTML = nodeHtmlAndClasses[i]["iH"];
    topNodes[i].className = nodeHtmlAndClasses[i]["cN"];
}
```

Search

Indexes and glossaries are nice, but search is everywhere. Search is an expected feature on the web, and it is often the first place someone goes when looking for an answer. Since so much of search in CHMs is handled by the CHM engine, we won't explore that here, but HTML5 search is another matter. Flare generates a sizable set of data files to support search in HTML5 output, so one of our examples will explore that.

Customizing search

Flare *skin* files let you define the appearance of a search field. When a user clicks the search button, Flare searches a set of data files called search chunks to return your results. Flare places these files in the output data folder, and you can use them to create customized output. Flare version 11 includes a search proxy that lets you embed a search field in your output. The next section describes how to do this.

Programmatic access to search in HTML5 output

As with TOC data files, there is a chunking scheme for search data, but the setup for search data files was not changed with version 9 as the TOC data files were. The index and glossary data files have also not changed. When you look in an output's data folder, you'll see XML and JavaScript files for the search data. Both files contain the same information. The only difference is that the JavaScript file places a wrapper function around the same XML that is in `Search.xml` (the `MadCap.Utilities.Xhr._FilePathToXmlStringMap.Add()` function, which is supplied by MadCap). This JavaScript wrapper function lets you read the XML from JavaScript code.

In the `Search.xml` and `Search.js` files, the `<urls>` element contains the data that will appear in search results. Each entry – the `<Url>` elements – has three attributes: `Source` (a relative link), `Title`, and `Abstract` (the text from a topic that appears in the search results). Example 6.7 shows a simple search data file.

Example 6.7 – XML markup for a search data file (`Search.xml`)

```xml
<?xml version="1.0" encoding="utf-8"?>
<index PreMerged="false" SearchType="Stem" NGramSize="1">
  <urls>
    <Url Source="../Content/Apples.htm" Title="Apples"
        Abstract="Apples Apples are tasty. " />
    <Url Source="../Content/Oranges.htm" Title="Oranges"
        Abstract="Oranges Oranges are orange and tasty. " />
  </urls>
  <ents>
    <stem n="appl" chunk="Search_Chunk1.xml" />
    <stem n="tasti" chunk="Search_Chunk1.xml" />
    <stem n="." chunk="Search_Chunk1.xml" />
    <stem n="orang" chunk="Search_Chunk1.xml" />
  </ents>
  <chunkfiles>
    <Url Source="Search_Chunk1.xml" />
  </chunkfiles>
</index>
```

Following the `<urls>` element is a list of stem elements contained by an `<ents>` element. Each of these entries has an n attribute and a `chunk` attribute. The `chunk` attribute points to the chunk file with more information. The n tag, for n-gram, contains parts of search terms – the stems.

The final list in the `Search.xml` and `Search.js` files is a list of chunk files.

The search chunk files (see Example 6.8) contain additional information about each stem listed in the Search.xml file. Each stem has a `<stem>` element that, as in the Search.xml or Search.js file, contains an n attribute for the stem string. Each `<stem>` element can contain multiple `<phr>` elements, one for each phrase that is associated with the stem. For example, the stem "page" may contain information about the phrases "page," "Page," and "pages." Each phrase has a list of `<ent>` tags, each of which points to an occurrence of the phrase.

Example 6.8 – XML markup for a search chunk data file (`Search_Chunk1.xml`)

```xml
<?xml version="1.0" encoding="utf-8"?>
<index>
 <stem n="appl">
  <phr n="Apples">
   <ent r="1000" t="0" w="1" />
   <ent r="102" t="0" w="3" />
   <ent r="3" t="0" w="5" />
  </phr>
 </stem>
 <stem n="tasti">
  <phr n="tasty">
   <ent r="3" t="0" w="7" />
   <ent r="3" t="1" w="9" />
  </phr>
 </stem>
 <stem n=".">
  <phr n=".">
   <ent r="3" t="0" w="8" />
   <ent r="3" t="1" w="10" />
  </phr>
 </stem>
 <stem n="orang">
  <phr n="Oranges">
   <ent r="1000" t="1" w="1" />
   <ent r="102" t="1" w="3" />
   <ent r="3" t="1" w="5" />
  </phr>
  <phr n="orange">
   <ent r="3" t="1" w="7" />
  </phr>
 </stem>
</index>
```

The `<ent>` tag contains three attributes. The first, `r`, is a weight. The second, `t`, indicates the topic. It is an integer that corresponds to the order of items in the urls section of the `Search.xml` and `Search.js` files. The last attribute is `w` and it indicates the position of the occurrence in the topic.

To grab an entry, the algorithm could first look for the stem in either the `Search.xml` or `Search.js` file. From there, the algorithm would look up the stem in the applicable chunk file. Once the stem is found, the algorithm would find the exact entry and then return the list of occurrences, relating the `t` attribute to the `<Url>` element in the `Search.xml` or `Search.js` file. The results could be ordered by `r`, and `t` could be used to open the topic to a particular location.

Search filters

There is a file type in Flare projects devoted to search, specifically *search filters* (`*.flsfs`). You can use search filters to define a filter to appear in the search field. The concepts you associate with a given filter are the only concepts that will appear in search results when a user selects the filter.

Example 6.9 – XML markup for a search filter

```
<?xml version="1.0" encoding="utf-8"?>
<CatapultSearchFilterSet>

 <SearchFilter Name="My Subset"></SearchFilter>

</CatapultSearchFilterSet>
```

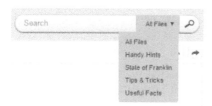

Figure 6.3 – Search filter in HTML5 output

Open search results for HTML5 output example

When a user clicks search in HTML5 output, a qualifier is appended to the URL. You can go straight to search results by creating that qualified URL yourself.

Example 6.10 – Simple input field to open a qualified URL for HTML5 search

```
<script>
  function open_help_search() {
    var uri = 'http://www.example.com/default.htm';
    var reference = encodeURIComponent(
        document.getElementsByName('help-search-text')[0].value);
    if (reference != "") {
        reference = '#search-' + reference;
    }
    window.open(uri + reference);
  }
</script>
<form name="help-search-form">
  <input type="text" name="help-search-text" />
  <button type="button" onclick="open_help_search()">Search help</button>
</form>
```

Batch testing search performance

For the last example in this chapter, let's look at a script to automate testing. The settings for search on a target file are mostly concerned with performance. Suppose you want to try out several dozen variations of the settings and monitor each for performance. Would you rather create one target, build it, test it, change something, build it, test it, and so on for hours? Or would you rather programmatically generate dozens of variations of a target file, programmatically generate batch files to build those targets, and while you are at it, programmatically open each output and measure the load time?

Example 6.11 does just that. It varies the values of six boolean variables (the variables assigned at the beginning of the Main() routine) to create 64 variations of a single target test file. We discuss how to generate batch files to build targets in Chapter 8.

Example 6.11 – C# tool to create multiple targets to test search settings

```csharp
using System;
using System.Collections.Generic;
using System.Linq;
using System.Xml.Linq;
using System.Text;

namespace TargetSearchVariations
{
  class Program
  {
    static void Main(string[] args)
    {
      Boolean includeStopWordsInSearch = false;
      String includeStopWordsInSearchString = "_stopWords_";
      Boolean excludeNonWordsFromSearch = false;
      String excludeNonWordsFromSearchString = "_nonWords_";
      Boolean excludeIndexEntriesFromSearch = false;
      String excludeIndexEntriesFromSearchString = "_indexEntries_";
      Boolean preMergeSearchDatabaseFile = false;
      String preMergeSearchDatabaseFileString = "_preMerge_";
      Boolean chunkLargeSearchDatabaseFiles = false;
      String chunkLargeSearchDatabaseFilesString = "_chunk_";
      Boolean enablePartialWordSearching = false;
      String enablePartialWordSearchingString = "_partialWord_";

      String baseTargetName = "target_";

      XDocument targetFile = XDocument.Load(args[0]);

      Char[] binaryValueText = { '0', '0', '0', '0', '0', '0' };

      for (int i = 0; i < 64; i++)
      {
        String binaryValueString = Convert.ToString(i, 2);
        for (int j = binaryValueString.Length; j < 6; j++)
        {
          binaryValueString = "0" + binaryValueString;
        }
        binaryValueText = binaryValueString.ToCharArray();

        includeStopWordsInSearch = binaryValueText[0].Equals('1');
        excludeNonWordsFromSearch = binaryValueText[1].Equals('1');
        excludeIndexEntriesFromSearch = binaryValueText[2].Equals('1');
        preMergeSearchDatabaseFile = binaryValueText[3].Equals('1');
        chunkLargeSearchDatabaseFiles = binaryValueText[4].Equals('1');
        enablePartialWordSearching = binaryValueText[5].Equals('1');

        String newTargetName = baseTargetName +
```

```
                    includeStopWordsInSearchString +
                    includeStopWordsInSearch.ToString().Substring(0, 1) +
                    excludeNonWordsFromSearchString +
                    excludeNonWordsFromSearch.ToString().Substring(0, 1) +
                    excludeIndexEntriesFromSearchString +
                    excludeIndexEntriesFromSearch.ToString().Substring(0, 1) +
                    preMergeSearchDatabaseFileString +
                    preMergeSearchDatabaseFile.ToString().Substring(0, 1) +
                    chunkLargeSearchDatabaseFilesString +
                    chunkLargeSearchDatabaseFiles.ToString().Substring(0, 1) +
                    enablePartialWordSearchingString +
                    enablePartialWordSearching.ToString().Substring(0, 1);

                targetFile.Root.SetAttributeValue("IncludeStopWords",
                    includeStopWordsInSearch.ToString());
                targetFile.Root.SetAttributeValue("ExcludeNonWords",
                    excludeNonWordsFromSearch.ToString());
                targetFile.Root.SetAttributeValue("ExcludeIndexEntriesFromSearch",
                    excludeIndexEntriesFromSearch.ToString());
                targetFile.Root.SetAttributeValue("PreMergeSearchDatabase",
                    preMergeSearchDatabaseFile.ToString());
                targetFile.Root.SetAttributeValue("ChunkSearchDatabaseIncludeStopWords",
                    chunkLargeSearchDatabaseFiles.ToString());
                targetFile.Root.SetAttributeValue("PartialWordSearch",
                    enablePartialWordSearching.ToString());

                targetFile.Save("C:\\TargetVariations\\" + newTargetName + ".fltar");
                Console.Out.WriteLine("Created: " + newTargetName);
            }
        }
    }
}
```

CHAPTER 7
JavaScript in Flare Topics, Master Pages, Snippets, and Skins

If you want to use JavaScript in an online output such as HTML5, you have two main options:

- Include your script in the Toolbar JavaScript through the Skin Editor
- Include your script in a topic, master page, or a snippet

Toolbar JavaScript through the Skin Editor

This is the path to take if you want the script to be available to the skin's chrome. This is the most global approach. If you want a JavaScript function available in your help system, this is the method to use. However, HTML5 tri-pane help uses iframes. So not everything is accessible from everywhere. For example, if you want to create a behavior in the search field, this is a good way to get it there. Toolbar JavaScript lets you define functions that can be called from buttons on the toolbar and the buttons themselves. But you don't have to stay within those bounds.

The Skin Editor has a tab called **Toolbar**. To the right of the **Toolbar** tab is a field that displays scripts and an **Edit** button. You can create and modify scripts using the **Edit** button, which opens an editing window. However, we don't recommend using this window for anything more than casual editing. You're better off using an editor that supports syntax highlighting, auto-completion, and the other features you expect from a good editor. If you use a different editor, just paste your code into the Skin Editor when you're done.

When you build your target, Flare places the Toolbar JavaScript in a file with this relative path:

```
Skins\Default\Scripts\Toolbar.js
```

The contents of this file are used exactly as you enter them. For example, if you place the following line in your Toolbar JavaScript, a popup with the text "Hello World!" will appear when you open the output.

```
alert('Hello World!');
```

If you will not be referring to a function externally (for example, by connecting it to a button), try to wrap the code in an anonymous function to keep the global namespace as clean as possible.

Here is a pattern you can use to wrap everything up into an anonymous function using a function expression. If you define your functions inside an anonymous function and only call those functions from within the anonymous function, those names won't conflict with other JavaScript code – either yours or MadCap's.

```
(function () {alert("Hello World!");}());
```

Unfortunately, an anonymous function won't work if you want to connect functions to buttons. Toolbar JavaScript is executed when the help output is opened, and functions placed in Toolbar JavaScript are accessible to toolbar buttons. So we need to define our Hello World function in the Toolbar JavaScript as follows:

```
function helloWorld() {alert("Hello World!");}
```

Connecting a Toolbar function to a button

To get a toolbar button to show that message takes the following steps. First, create a new button in the skin by opening the Skin Editor, then clicking the **add** icon, which is on the **Toolbar** tab under **Toolbar Buttons**. Enter a name in the screen that appears and click **OK**. This makes the button available, but doesn't add any functionality, yet. Highlight the new button under **Available** and use the arrow to move it to the **Selected** list. From there you can determine the order of buttons.

Now that the button is in, you can adjust the look and add a click behavior. To do that, go to the **Styles** tab and scroll to the **Toolbar Button** section. Select the **new** button. The **Properties** section lists many of the CSS properties, which you will probably want to adjust. The click behavior is defined under **Event** at the bottom of the list. To create the "Hello World!" message, enter the following string for the event:

```
helloWorld();
```

When you build the output, you will see a new button in the toolbar that displays "Hello World!" when clicked.

There are lots of possibilities here. But, the basic pattern remains the same. You define functions, place them in the Toolbar JavaScript, and call them from buttons.

Connecting Toolbar functions to other actions

You can connect Toolbar functions to more than buttons. Toolbar JavaScript is just a means of defining a script to be included in the output. Remember that when we first added the "Hello

World!" function to the Toolbar JavaScript, we didn't wire it up to a button. You have other options. For example, if you add the code in Example 7.1 to the Toolbar JavaScript, it will add functionality to the search field that clears any text in the field when you press the escape key.

Example 7.1 – JavaScript to add keyup behavior for the search field

```
window.onload = function() {
  $("#search-field").keyup(function (e) {
    if (e.keyCode == 27) { $("#search-field").val(""); }
  });
}
```

There are two other ways to add JavaScript to appear in skin-related files. One way is to insert markup, scripts, etc., after you generate your output files. When you do this, keep in mind you have to do it every time you rebuild. The other way is to modify the skin templates in the installation folder for the Flare application. When you do this, the change takes effect for any subsequent build that uses the modified templates. If you take this approach, back up the templates first.

Scripts in topics, master pages, and snippets

You don't have to put your scripts in the Toolbar JavaScript. You can place JavaScript files in the `Content` folder and point to those files using the `src` attribute on the `<script>` tag in topics, master pages, or snippets. For that matter, you can reference external sources just as you would in any HTML page.

When you place a script tag in a topic, the topic is the only thing that gets the script. The other topics don't get it, and the skin doesn't get it. The topic is either shown as its own page or in the content iframe in the skin. If only one topic needs a script, that may be the best place to put it.

When you place a script in a master page, every topic to which the master page is applied gets it. When you place a script in a snippet, any topic, snippet, or master page that includes the snippet gets the script. If the snippet is included as just text, the script doesn't make it into the output.

CHAPTER 8
Document Automation and Batch Files

This chapter discusses techniques for automating Flare target builds, using help subsystems, and writing batch files to facilitate document automation.

Batch files

You can create batch targets for a project. From the Batch Target Editor, you can select actions and schedule tasks. To generate output from Flare targets without using the Flare user interface, you can use the command line executable `madbuild.exe`, which is located in the `Flare.app` folder of your Flare installation.

If you do not specify a target when you call `madbuild.exe`, Flare will build every target in the project. You can enable logging with `-log true` appended to the end of the command. Log files appear in the project in the `Reports` folder.

The primary benefit of using batch processing is scheduling. Large API and database references can take hours to build. Scheduling builds means you don't have to remember to manually kick off builds, and you don't need to be present when the build starts. You can set up other tasks as batch processes, too, such as the following:

- Scheduling post-build processing tasks.
- Scheduling parallel builds for outputs with multiple subsystems.
- Automating source control actions. For example, Microsoft's Team Foundation Server (TFS) provides the `tf.exe` executable, and other source code systems provide similar executables.
- Post-processing PDF output. For example, Adobe Acrobat supports an API that can be used by command-line utilities.
- Copying the output or the source to an archive.
- Copying a different version of a stylesheet into the output.
- Copying external files into the project folder structure.
- Adjusting the skin CSS beyond what is possible in the Skin Editor.

Creating a manageable template for batch files

Nobody wants to copy a long command and change just one item over and over again. One way around that in batch files is to use variables for commonly used items such as the paths for the project folder or the Flare.app folder. Example 8.1 creates variables for these two folders and for the project file, then calls madbuild.exe using those variable for the arguments passed in.

Example 8.1 – Batch file to run madbuild.exe

```
rem set these for your project
rem set the target down below
set prjfldr="C:\flare-projects\example"
set prjnm="example.flprj"
set flarefldr="C:\Program Files (x86)\MadCap Software\MadCap Flare V10\Flare.app"
cd "%flarefldr%"
rem set target here and add other lines as necessary
madbuild.exe -project "%prjfldr%\%prjnm%" -target "build-for-import.fltar"
```

If you use source control, you can check files in and out using a batch process. For example, you can use the Team Foundation Server tf.exe executable. Example 8.2 shows a batch file that checks out files using tf.exe, runs a build, and then checks the files back in. When you check items in and out this way, make sure you don't include items, such as reports, analyzer data, and outputs, that don't belong in source control.

To use the batch file in Example 8.1, adjust the variables, including the locations of the tf.exe file and the Flare.app folder, to match your configuration. If you use a different source control solution, such as Subversion, you will need to change the tf.exe calls to match the syntax for that solution.

Example 8.2 – Batch file to check out, process, and check in files

```
rem set these for your project
rem set the target down below
set projectfldr="C:\flare-projects\example"
set project="%projectfldr%\Project"
set projectfile="%projectfldr%\example.flprj"
set vsfldr="C:\Program Files (x86)\Microsoft Visual Studio 11.0\Common7\IDE"
set flare="C:\Program Files (x86)\MadCap Software\MadCap Flare V10\Flare.app"
cd "%vsfldr%"
tf.exe checkout "%projectfldr%\Content" /recursive /noprompt
tf.exe checkout "%project%\Advanced" /recursive /noprompt
tf.exe checkout "%project%\ConditionTagSets" /recursive /noprompt
tf.exe checkout "%project%\Destinations" /recursive /noprompt
tf.exe checkout "%project%\Glossaries" /recursive /noprompt
tf.exe checkout "%project%\Imports" /recursive /noprompt
tf.exe checkout "%project%\Skins" /recursive /noprompt
tf.exe checkout "%project%\Targets" /recursive /noprompt
tf.exe checkout "%project%\TOCs" /recursive /noprompt
tf.exe checkout "%project%\VariableSets" /recursive /noprompt
cd "%flare%"
rem set target here and add other lines as necessary
madbuild.exe -project "%projectfile%" -target "build-for-import.fltar"
cd "%vsfldr%"
tf.exe checkin "%projectfldr%\Content" /recursive /noprompt
tf.exe checkin "%project%\Advanced" /recursive /noprompt
tf.exe checkin "%project%\ConditionTagSets" /recursive /noprompt
tf.exe checkin "%project%\Destinations" /recursive /noprompt
tf.exe checkin "%project%\Glossaries" /recursive /noprompt
tf.exe checkin "%project%\Imports" /recursive /noprompt
tf.exe checkin "%project%\Skins" /recursive /noprompt
tf.exe checkin "%project%\Targets" /recursive /noprompt
tf.exe checkin "%project%\TOCs" /recursive /noprompt
tf.exe checkin "%project%\VariableSets" /recursive /noprompt
```

Automating imports

Flare targets have an Auto-Sync feature that lets you run imports as an early step in generating output. Auto-Sync can be enabled or disabled for a target from the Target Editor. The setting is on the **General** tab, under **Auto-Sync: Disable auto-sync of all import files.** The value of the `DisableAutoSync` attribute on the `<CatapultTarget>` element will be set to `true` when this box is checked and `false` when it is unchecked.

When you enable Auto-Sync on a target, you can tell Flare to run an import before generating output by selecting the **Auto-reimport before "Generate Output"** on the **Source Project** tab. The markup for this option in the import file is the attribute/value pair `AutoSync="true"` on the `<CatapultProjectImport>` element. The option on the target is **Disable auto-sync of all import files**, and the markup on the target is `DisableAutoSync="false"`

When a target is built with `madbuild.exe`, Auto-Sync will be run if it has been enabled. You can automate imports by running a build for a target with Auto-Sync enabled. To schedule a build, create a batch file with the appropriate build commands and schedule a call to the batch file with Windows Scheduler.

Building subsystems in parallel

Building in parallel means you first build the subsystems at the same time (i.e., in parallel), and then after the subsystems are built, you build an output that incorporates those subsystems. This can be useful when you have many targets. You can run several commands at the same time from one batch file using the `start` command. The `start` command doesn't wait for the command it starts to complete; it returns immediately after starting the `madbuild.exe` command. This means, for example, that in Example 8.3, the two `madbuild.exe` commands will run in parallel.

Example 8.3 – Parallel build using the `start` command

```
start madbuild.exe -project "C:\Example\Example.flprj" -target "Example1.fltar"
start madbuild.exe -project "C:\Example\Example.flprj" -target "Example2.fltar"
```

When you have a parallel batch process like this, all the targets must be built before you try to build the complete system. To ensure that this happens, you can monitor execution from an external process or create a process to check that the targets have been built.

Batch file generator utility

Figure 8.1 shows a Windows Forms application that lets you generate a batch file by filling in a one-page form. The form lets you specify options, point to a folder, and generate a batch file with `madbuild.exe` commands for each target in the folder. You can also have the application build a separate batch file for each `madbuild.exe` command.

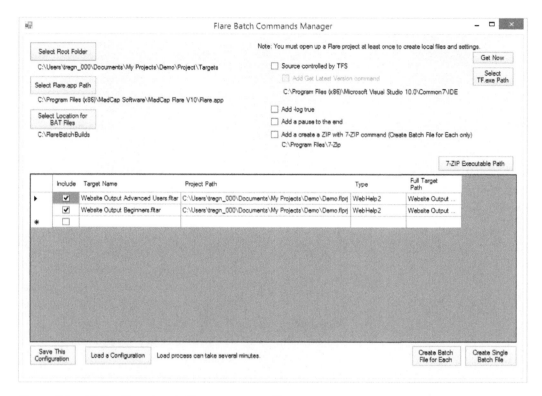

Figure 8.1 – UI for Flare batch file generator utility

The source code for the utility can be found in Appendix A, *Batch File Generator Utility* (p. 139), and also can be downloaded from the companion website (http://xmlpress.net/flare/examples).

Using the utility

Table 8.1 describes the input fields used by the batch file generator utility. Start with the three buttons to the left-hand side of the interface (see Figure 8.1), and then fill in the form fields as described in the table. The buttons at the bottom of the form let you save or load a configuration and create single or multiple batch files.

Table 8.1 – Batch commands manager input fields

Field	Explanation
Select Root Folder	Opens a file dialog to select the folder that contains your targets. The grid refreshes when you change this path.
Select `Flare.app` Path	Opens a file dialog to select the folder that contains `madbuild.exe`.
Select Location for BAT Files	Opens a file dialog to select the folder where the utility will place the generated batch files.
Grid	The grid contains these columns: Include: If checked, the utility will include a line in the batch file for the target. If not checked, the utility will still generate the line, but it will make the line a comment. Target Name: The filename for the target. Project Path: The path for the project, including the project filename. Type: The target type, such as PDF. Full Target Path: The target path, including the filename.
Source controlled by TFS	Checkbox to indicate whether the project is maintained using Team Foundation Server. This check box enables the **Add Get latest Version command** checkbox.
Add Get latest Version command	Checkbox to add code at the beginning of each batch file that will ask TFS to update the files to the latest version
Add `-log true`	Checkbox to add `-log true` to the build command lines.
Add a pause to the end	Checkbox to add a command line for Pause at the end of each batch file.
Add a create a ZIP with 7-ZIP command	Checkbox to add lines near the end of the batch files to place the output in a ZIP file using the `7-ZIP` command.

Field	Explanation
7-ZIP Executable Path	Opens a file dialog to select the folder that contains the `7-ZIP` command.
Save This Configuration	Saves the current configuration of the Flare Batch Commands Manager as an XML file. The configuration file saves all of the information entered in the form for later use.
Load a Configuration	Loads a previously saved configuration. Opens a file dialog to select the XML configuration file. Note: A configuration with many targets can take a while to load.
Create Batch File for Each	Generates separate batch files for every target in the grid.
Create Single Batch File	Generates a single batch file with one build line for each target.

CHAPTER 9
Connecting Applications to Flare Help Outputs

The Flare online and application help have topics that describe how to use context sensitive help with DotNet Help, HTML Help, HTML5, WebHelp, and WebHelp Mobile. In this chapter, we discuss some examples that go beyond those topics.

HTML5 tri-pane help

One of the biggest differences between WebHelp and HTML5 is the move from HTML framesets, which are used in WebHelp, to iframes, which are used in HTML5 tri-pane help. The HTML5 standard does not support framesets.

Because of this, HTML5 URLs require a different syntax. For example, to specify a topic with WebHelp, you append a query string that identifies the topic to the main part of the URL using a question mark. But with HTML5, you create the URL for a specific topic by appending a fragment that identifies the topic using a hash symbol ("#").

HTML5 URL examples

Here are some examples of how to build URLs to HTML5 help topics. They use the MadCap Flare online help for version 10 as the base.

Here is the URL for the default page for MadCap Flare version 10 online help. This is the expected main landing page for the help system:

```
http://webhelp.madcapsoftware.com/flare10/Default.htm
```

Note: To keep the URLs from being too long, we will use the variable *BASEURL* to represent the base URL for this help system: `http://webhelp.madcapsoftware.com/flare10`.

Here is the URL for the Welcome topic in MadCap Flare version 10 online help. Since the Welcome topic is the default topic for the system (as of this writing), that topic is displayed when you go to the main landing page.

When a user clicks the Welcome topic in the Contents in the navigation area, the same topic is still displayed, but the URL will change to the following:

`BASEURL/Default.htm#Introduction_Topics/Welcome_Flare.htm%3FTocPath%3D_____1`

Here is the URL for a topic to be displayed without the TOC path information:

`BASEURL/Default.htm#Introduction_Topics/Welcome_Flare.htm`

Here is the URL for the welcome page as a standalone topic:

`BASEURL/Content/Introduction_Topics/Welcome_Flare.htm`

Here is the URL for a search for the word "welcome":

`BASEURL/Default.htm#search-welcome`

By default the startup topic (the Welcome topic in our example) is the first topic in the TOC, but users can configure the system to use a different target as the startup topic. This, as well as other syntax possibilities, is described in the MadCap Flare help.[1]

When a topic is viewed as a standalone topic, all the extra chrome – navigation tabs, logo, search box, etc. – from the HTML5 skin is not shown. Thus, you can create a help system that only shows standalone topics and never shows the skin.

When a user types a word into the search field and clicks the search icon, the URL changes to the search syntax. So you don't need to show the search field to use search or show search results. For example, if you type the following into your browser, Flare will give you search results for the word "troubleshooting":

`BASEURL/Default.htm#search-troubleshooting`

With this knowledge in hand, you can build URLs that show specific topics in different contexts and even create a custom interface to open search (We show an example of this in Example 6.10, "Simple input field to open a qualified URL for HTML5 search" (p. 76)). For example, here is a URL that tells Flare to show the first pick in a search instead of showing all results:

`BASEURL/Default.htm#&searchQuery=welcome&firstPick=true`

---

[1] http://webhelp.madcapsoftware.com/flare10/Default.htm#Targets/More_About_Targets/Specifying_the_Startup_Topic.htm

HTML5 JavaScript helpers

HTML5 output includes JavaScript helper functions that you can use as aids in displaying help. For example, instead of building URLs to display topics, you can use the JavaScript function `MadCap.OpenHelp`. The JavaScript helpers are located in the output file called `csh.js`. Code comments in that file describe usage.

The `MadCap.OpenHelp` function (see Example 9.1) takes five parameters (`null` means default):

- **id:** The CSH (context sensitive help) ID or a path to a topic.
- **skinName:** The name of the skin file without the extension.
- **searchQuery:** The search string, which is the same as what you would enter in the search field of a URL.
- **firstPick:** If `true` and searchQuery is not `null`, open the first item in the search results. If there are no results, the topic pane will be empty.
- **pathToHelpSystem:** Use this value for webHelpPath instead of building the URL. This parameter isn't described in the help, but it is in the method signature.

Example 9.1 shows the JavaScript for a button to open the help topic with the ID `SpecialTopic`.

Example 9.1 – JavaScript button that calls `MadCap.OpenHelp`

```
<input type="button" value="Show me the help!"
       onclick="MadCap.OpenHelp('SpecialTopic', null, null, null );" />
```

IDs let you create topic aliases in a Flare project and map those aliases to specific topics. You give a set of IDs to a developer once, and the developer codes the application to open the help using the appropriate ID either in a URL or as a parameter passed to the `MadCap.OpenHelp` function. You can also use a topic path for the ID parameter.

For example, suppose you have a topic called `Oranges.htm`, and you want to give developers access to that help topic in their applications. To do this, you edit the alias (`AliasFile.flali`) and header (`HeaderFile.h`) files using the Alias Editor, which you can do in Flare by going to **Project → Advanced → CSH**. Opening the alias file will bring up the Alias Editor.[2]

---

[2] Opening the header file will bring up the Text Editor, but normally you don't do that. Instead, standard practice is to manage both the header file and the alias file through the Alias Editor.

In the Alias Editor, you can assign an identifier to `Oranges.htm` that has two forms: a character string ID name, such as `Fruit1`, and a numeric ID value, such as `1000`. The Alias Editor places the output alias information in the `Data` folder in two forms: as an XML file and, via a transformation, as a JavaScript file in. The JavaScript file is how the mappings from topic name to ID get loaded into an object.

The alias file associates a topic path – in our example, `Oranges.htm` – with an ID name, but it does not contain the numeric ID value. You specify the ID name in the **Identifier** column in the Alias Editor. If you specify a skin, it will appear as an attribute on the `<Map>` element. See Example 9.2 for a sample alias file.

Example 9.2 – Sample alias file (`AliasFile.flali`)

```
<?xml version="1.0" encoding="utf-8"?>
<CatapultAliasFile>
    <Map Name="Fruit1" Link="/Content/Apples.htm" />
    <Map Name="Fruit2" Link="/Content/Oranges.htm" />
</CatapultAliasFile>
```

You specify the numeric ID value in the **Value** column in the Alias Editor. This numeric identifier is kept in the header file, `HeaderFile.h`, and the value must be unique. The header file only includes the name and numeric identifier; it does not refer to the file name. This means the header file does not change if you change the path for a particular topic. Developers can use the header file in their programs to give each ID value an easy-to-use ID name (see Example 9.3).

Example 9.3 – Sample header file (`HeaderFile.h`)

```
#define Fruit1 1
#define Fruit2 1000
```

You publish the help system and tell the developer to pass the ID value `1000` to `MadCap.OpenHelp` from whatever event is used to execute the function that uses this help topic. The developer codes this and from that point on, whenever that event occurs, `Oranges.htm` is displayed. The developer can pass in the ID value directly or include the header file in the program and refer to the ID name.

Later, if you no longer want to show `Oranges.htm` for that event, the developer doesn't need to do anything. You simply change the mapping for ID name `Fruit1` to point to, for example,

Apples.htm. Once the target help system is rebuilt and published, the application will show Apples.htm when ID name Fruit1 or ID value 1000 is used. There is no code churn, and the developer doesn't have to invest any time in the process.

Java example

You can open a browser-based help system from a Java application. One approach for doing this is to build the URL in the Java application and use a Java library to open a browser with that URL.

Example 9.4 shows some snippets of a Java application that do just that.

Example 9.4 – Java methods for opening URLs

```
public void showHelp(String key) {
   URI uri;
   try {
      uri = new URI(
"http://tregner.com/example-projects/java-app-help/java-app-help.htm#cshid="
            + key);
      openWebpage(uri);
   } catch (URISyntaxException ex) {
      Logger.getLogger(
            JFrameHtml5Example.class.getName()).log(Level.SEVERE, null,
            ex);
   }
}

public static void openWebpage(URI uri) {
   Desktop desktop =
         Desktop.isDesktopSupported() ? Desktop.getDesktop() : null;
   if (desktop != null && desktop.isSupported(Desktop.Action.BROWSE)) {
      try {
         desktop.browse(uri);
      } catch (Exception e) {
         e.printStackTrace();
      }
   }
}

public static void openWebpage(URL url) {
   try {
      openWebpage(url.toURI());
   } catch (URISyntaxException e) {
      e.printStackTrace();
   }
}
```

For Java programs, we need to use a variation on the header file called a `*.properties` file. You can export a `*.properties` file from the **Tools** ribbon in Flare. The `*.h` extension and syntax is based on a convention used in C and C++, but the `*.properties` convention is more commonly used in Java, and there are library methods available to map properties defined in a `*.properties` file to Java Properties. As with the header file, the `*.properties` file maps ID names to numeric ID values.

For more information, see this Java Tutorial at docs.oracle.com.[3]

Example 9.5 – Sample properties file (`HeaderFile.properties`)

```
Fruit1 = 5
Fruit2 = 1000
```

Visual Basic 6

As with Java, a Visual Basic 6 application can use a file to load topic and alias information. In the case of Visual Basic 6, the file extension is `*.bas` and exporting header information to `*.bas` is an option on the **Export Header Files** screen.

Visual Basic .NET and C#

We usually recommend selecting a convention and sticking with it, but `*.bas` is a legacy format, and it is probably only appropriate to use that extension with Visual Basic 6. For Visual Basic .NET and C#, the header file should meet your needs. The technique is similar to Example 9.4. And Flare includes a DotNet Help target type that is designed to work with .NET applications to present dynamic help.

Pascal

Flare provides an option to export a Delphi Pascal `*.inc` formatted header file.

PHP

MadCap Support has posted a PHP example on their support site.[4]

---

[3] http://docs.oracle.com/javase/tutorial/essential/environment/properties.html
[4] http://kb.madcapsoftware.com/#Flare/Context_Sensitive_Help/CSH1003F_-_Context_Senstive_Help_call_using_PHP.htm

WebHelp

WebHelp is a precursor to HTML5, and in fact, the HTML5 target type is alternately referred to as WebHelp 2.0. If you look at the source files for the skins, you will see WebHelp2, not HTML5. You can find those files in the following location:

```
C:\Program Files (x86)\MadCap Software\MadCap Flare V10\Flare.app\Resources
```

However, even with the release of HTML5 output, WebHelp is not a legacy format, yet. If you still use WebHelp, consult the Flare online and application help for a description of the differences.

CHM and HTML Help

Over the years, security concerns and newer versions of Windows have made the CHM format all but obsolete. However, for some organizations there remain business reasons to create CHM or HTML Help outputs.

You can use the HTML Help engine to open a CHM, and you can even point to specific topics. Here is an example:

```
HWND HtmlHelp (Window(), "C:\Examples\Example.chm", HH_DISPLAY_TOPIC,
               "SpecialTopic.htm");
```

DotNet Help

MadCap provides a redistributable help viewer and a sample .NET application that you can download from the MadCap website to help you connect a .NET application to DotNet Help. The MadPak suite, including Flare, uses DotNet Help as the engine for dynamic help. You can find the redistributable help viewer here:

```
http://www.madcapsoftware.com/downloads/redistributables.aspx
```

You can use the redistributable help viewer in the following three ways:

- Run the redistributable help viewer from the command line
- Use library API class HelpViewerClient to create a client
- Use library API class HelpViewerEmbeddedClient to create embedded help

Embedded help, facilitated by the HelpViewerEmbeddedClient API, is more tightly coupled and requires less coding for behaviors such as assigning an action to the **F1** key or dynamically rendering topics. On the other hand, the HelpViewClient API reduces coupling at the cost of leaving behaviors such as **F1** to be coded by you or your developers.

Eclipse Help

Eclipse Help was added to Flare as a target type in version 10. This is a nice addition for users of Eclipse or Eclipse-derived IDEs. Deploying the output uses the same process as deploying any Eclipse help. Kepler is the required version of Eclipse, which (as of this writing) is the most recently released version.

CHAPTER 10
Flare Application and Flare Plugins

As we have seen, the open XML project structure of MadCap Flare provides many ways to manipulate projects, topics, and even outputs. MadCap introduced a plugin API for Flare in version 9 that lets you customize the Flare user interface, too. However, the tools used to customize the interface are different. Flare is a *.NET* application that uses *DLLs* (dynamic-link libraries).

With version 10, Flare is installed by default in the following folder:

`C:\Program Files (x86)\MadCap Software\MadCap Flare V10`

Drilling into the subfolder `Flare.app`, you'll see DLLs and a handful of executables. The most important executable is, of course, `Flare.exe`. If you plan to build targets from outside of Flare, you'll also need `madbuild.exe`, which we discussed in Chapter 8. There are other executables included in Flare, but these two are enough for now.

In addition, there are dozens of *assemblies*, which are used by the executables. Assemblies encapsulate and organize code and resources for an application, making it possible for MadCap to issue patches that update only a single file. A .NET developer creates executables and assemblies (DLLs) by building projects – usually Visual Studio projects for a CLR (common language runtime) language such as C# or Visual Basic .NET. These executables and assemblies can be used by other executables on a Microsoft operating system.

Because code can be packaged in assemblies and executables, .NET applications are extensible. MadCap chose to use a plugin API to make Flare extensible. A Flare plugin is an assembly and associated resources, such as images or other files, that conform to certain rules. In particular, a Flare plugin must implement the interface defined in `B3.PluginAPIKit.dll`, an assembly that is located in the same folder as `Flare.exe`.

MadCap defined this interface to ensure that every Flare plugin follows the same rules and has the same properties. When you create a plugin, you add a reference to `B3.PluginAPIKit.dll`, and Visual Studio provides visibility into that assembly for your project.

Creating a plugin project

Let's get started with a project. You don't have to own a license for a full version of Visual Studio to do this. You can download Visual Studio Express, a free application that lets you create C# projects and, as of this writing, includes enough functionality to code a Flare plugin. As long as your computer meets the minimum system requirements for Visual Studio, you can create a plugin with no investment in a development environment. The only cost is your time learning to code for .NET.

Let's assume you have installed Visual Studio Express for Windows Desktop.[1] The next step is to create a new project. Select **File → New Project**. From the **New Project** screen, select **Visual C# → Class Library**. Enter `FirstPlugin` for the name and click **OK** (see Figure 10.1).

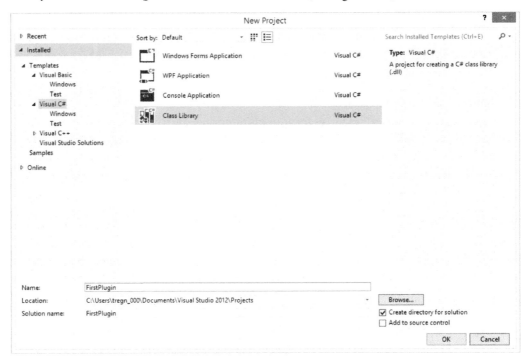

Figure 10.1 – **New Project** window in Visual Studio

---

[1] http://www.visualstudio.com/en-us/downloads/download-visual-studio-vs

The project will appear in Visual Studio. Open the Solution Explorer, if it isn't already open, by selecting **View → Solution Explorer**. The project is contained in a *solution*, a container for projects and other items. The project contains properties, references, and a C# class file (`Class1.cs`). If you double-click the class file in the Solution Explorer, the file will open in an editor. As you can see in Example 10.1, there won't be much there. Our job is to change that.

Example 10.1 – C# class file generated by Visual Studio (slightly adjusted)

```
using System;
using System.Collections.Generic;
using System.Linq;
using System.Text;
using System.Threading.Tasks;

namespace FirstPlugin
{
    public class Class1
    {
    }
}
```

But first, we need to make sure we are using the correct version of Microsoft's .NET framework. As of Flare version 10, the target framework is version 4.0. To change the target framework, double-click **Properties** for the project in Solution Explorer, then from the **Application** page select ".NET Framework 4" in the **Target framework** drop down (see Figure 10.2).

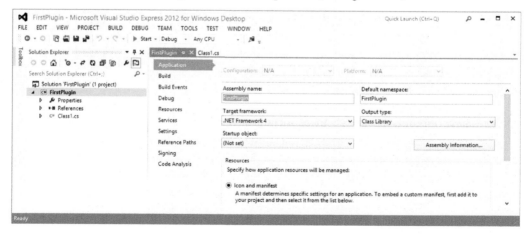

Figure 10.2 – Selecting the target framework in Solution Explorer properties

Now, we need to add a reference to B3.PluginAPIKit.dll. From Solution Explorer, expand the **References** drop down in the project. You should see a list. Eventually, we will add more than B3.PluginAPIKit.dll, but for now we will just add that one. Right-click **References** and select **Add Reference**. The window that appears (References Manager) gives you several ways to find an assembly. We want to browse to the folder where B3.PluginAPIKit.dll is located, so click **Browse** and find the assembly. In a default install, you will find B3.PluginAPIKit.dll here:

C:\Program Files (x86)\MadCap Software\MadCap Flare V10\Flare.app

Once you find it, click **Add** and then **OK**. The assembly will appear in the list of references (see Figure 10.3).

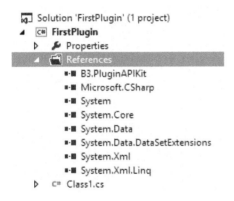

Figure 10.3 – Initial references in the project and the newly added B3.PluginAPIKit reference

We will need more references than just B3.PluginAPIKit.dll. Usually, you add references only as needed, but in this case, let's add all the references we need for this example. Figure 10.4 shows the complete list of references you need.

Some of these resources are already there, including B3.PluginAPIKit, as you can see in Figure 10.3. You can add the others by bringing up the Reference Manager window again. But this time, go to **Assemblies** and multi-select the items. All of these assemblies are from Microsoft. We will use them in our plugin to create custom menus and ribbons in the Flare user interface and to display those menus and ribbons via a Windows Form.

- B3.PluginAPIKit
- Microsoft.CSharp
- PresentationCore
- PresentationFramework
- System
- System.Core
- System.Data
- System.Data.DataSetExtensions
- System.Drawing
- System.Windows.Forms
- System.Xaml
- System.Xml
- System.Xml.Linq
- WindowsBase

Figure 10.4 – References needed for Flare plugins

You may know that before Flare version 8, you didn't have the option to switch between the classic menu interface and the ribbon interface; you could only use the classic menus. However, Flare now supports both styles, so you can choose either style or both. However, if you want to support both, you'll need to create two variations for your plugin UI.

Now, let's change the name of the class file to `FirstPlugin.cs`. Right-click the `Class1.cs` file, select **rename**, and change the name. When you press **enter**, you will see this dialog:

```
--------------------------
Microsoft Visual Studio
--------------------------
You are renaming a file. Would you also like to perform a rename in this
project of all references to the code element 'Class1'?
--------------------------
Yes   No
--------------------------
```

Click **Yes**. This will adjust the class name in the file. When you do that, an asterisk will appear on the tab for the Editor indicating there are unsaved changes. Go ahead and click **Save**.

We want to refer to some things inside these assemblies using a shortcut. So, edit the class file so it looks like Example 10.2:

Example 10.2 – C# `Class1.cs` file with assembly references added

```
using System;
using System.Collections.Generic;
using System.Linq;
using System.Text;
using System.Windows.Forms;
using System.Windows.Input;
using B3.PluginAPIKit;

namespace FirstPlugin
{
    public class Class1
    {
    }
}
```

These additions, which are commonly used in day-to-day .NET programming, enable us to use shorter names for items in those namespaces. Without those `using` statements, we would have to fully qualify names. For example, we will implement the interface using the `IPlugin` class in `B3.PluginAPIKit`. Without the `using` statement we would have to type the following:

```
public class FirstPlugin : B3.PluginAPIKit.IPlugin
```

But with the `using` statement, we can just type the following:

```
public class FirstPlugin : IPlugin
```

That may not seem like a big deal. But with lengthy namespaces and paths, it can make a big difference in the readability and writability of your code. You should also see IntelliSense kick in when you start typing `IPlugin` (see Figure 10.5) and when you enter the `using` statements.

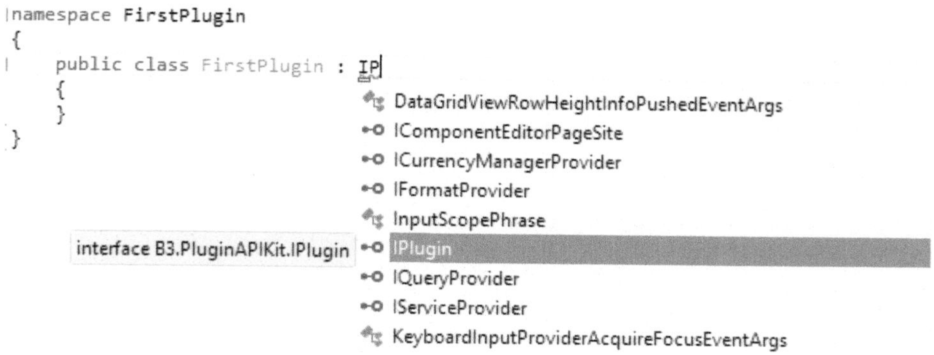

Flare Application and Flare Plugins   105

```
namespace FirstPlugin
{
    public class FirstPlugin : IP|
    {
    }
}
```

- DataGridViewRowHeightInfoPushedEventArgs
- IComponentEditorPageSite
- ICurrencyManagerProvider
- IFormatProvider
- InputScopePhrase
- **IPlugin**   interface B3.PluginAPIKit.IPlugin
- IQueryProvider
- IServiceProvider
- KeyboardInputProviderAcquireFocusEventArgs

Figure 10.5 – `IPlugin` appears in IntelliSense

You can press **tab** and **enter** to use autocomplete. This can be tricky if you are using the Visual Studio 2012 UI. There is a little floating drop down that appears when you place the cursor on `IPlugin` (see Figure 10.6). Sometimes the drop down is just a little rectangle under the first letter. It may be hard to get this to open, but as we'll see, it's worth the trouble.

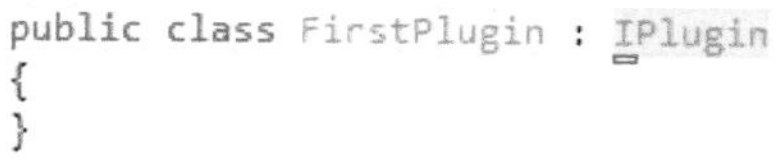

```
public class FirstPlugin : IPlugin
{
}
```

Figure 10.6 – Difficult-to-notice drop down

Open the drop-down menu and select **Implement interface 'IPlugin'** (see Figure 10.7).

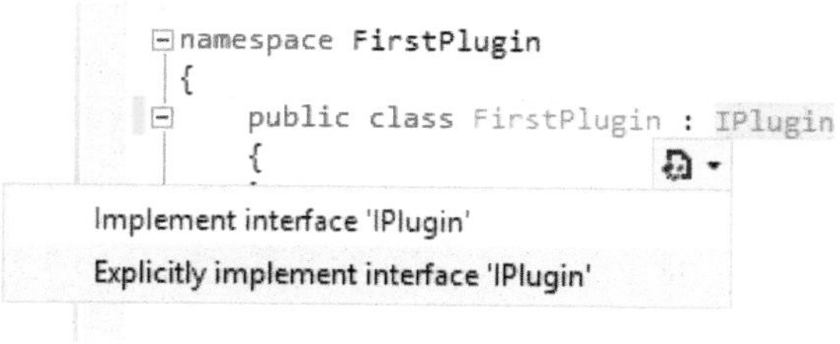

```
namespace FirstPlugin
{
    public class FirstPlugin : IPlugin
    {
```
Implement interface 'IPlugin'
Explicitly implement interface 'IPlugin'

Figure 10.7 – Drop down to implement the interface

This will add stubbed-out interface definitions from `B3.PluginAPIKit` to your class. Example 10.3 shows the result.

Example 10.3 – Stubbed out C# implementation of IPlugin interface (spacing adjusted)

```
1  using System;
2  using System.Collections.Generic;
3  using System.Linq;
4  using System.Xml.Linq;
5  using System.Text;
6  using System.Windows.Forms;
7  using System.Windows.Input;
8  using B3.PluginAPIKit;
9
10 namespace FirstPlugin
11 {
12     public class FirstPlugin : IPlugin
13     {
14         void IPlugin.Execute()
15         { throw new NotImplementedException(); }
16
17         string IPlugin.GetAuthor()
18         { throw new NotImplementedException(); }
19
20         string IPlugin.GetDescription()
21         { throw new NotImplementedException(); }
22
23         string IPlugin.GetName()
24         { throw new NotImplementedException(); }
25
26         string IPlugin.GetVersion()
27         { throw new NotImplementedException(); }
28
29         void IPlugin.Initialize(IHost host)
30         { throw new NotImplementedException(); }
31
32         bool IPlugin.IsActivated
33         { get { throw new NotImplementedException(); } }
34
35         void IPlugin.Stop()
36         { throw new NotImplementedException(); }
37     }
38 }
39
```

Example 10.3 contains just the stubs for the interface; there isn't any code, yet. In other words, we may have installed a sink with a faucet and knobs, but it isn't hooked up to any pipes, yet. Providing that code – the plumbing – is our job. Now, let's create some instance properties to be used by the plugin. At the beginning of the class (before line 14 in Example 10.3), add the following lines, which define four instance properties:

```
private IHost mHost;
private bool mActivated;
private INavContext mNav;
private IEditorContext mEditorContext;
```

If you want to explore one of these items, you can right click the data type (`IHost`, `bool`, `INavContext`, or `IEditorContext`) and select **Go To Definition**. Visual Studio will show you some more information.

MadCap does not provide any code comments for `B3.PluginAPIKit.dll`. Otherwise, there would be more detailed IntelliSense for the API. However, MadCap has API documentation for versions 9 and 10 that you can download as a ZIP file. That documentation describes the `IHost`, `INavContext`, and `IEditorContext` data types, and the `bool` type is `System.Boolean`, which is in `mscorlib.dll` (Microsoft's core library). We suggest you download the API documentation[2] and have it available as you work.

Note: The API functionality differs significantly depending on the Flare version. Prior to version 9, there was no plugin API at all. Version 9 added API functionality, but does not support manipulation of the DOM (Document Object Model), which was added with version 10. The first example here will work with version 9 or version 10, but examples that manipulate the DOM, such as Example 10.9, will not work with version 9.

Let's fill in some more details for the implementation. Each of the stubbed out methods should do something other than throw a `NotImplementedException`, so let's implement them now by editing the stubs in our class.

Initialize(IHost host)

Although this method isn't first in the file, it is key because it's the hook into everything else. This is the only method with a parameter. Flare will pass the parameter `host` to this method, and we'll get everything else based on it. But in order to access this information elsewhere, we assign the value of `host` to the instance property `mhost`, which we declared earlier. This takes care of one of the four instance properties.

```
public void Initialize(IHost host)
{
  mHost = host;
}
```

---

[2] http://www.madcapsoftware.com/downloads/redistributables.aspx

Execute()

Minimally we need to assign the other three instance properties, two of which are derived from host. When Execute() is called, the plugin will register as activated and the context for the navigation and editor will be provided. We assign these to instance properties for the same reason we assigned host to mhost – to access them elsewhere. Execute() is called when the user activates the plugin from Flare.

```
public void Execute()
{
  mActivated = true;
  mNav = mHost.GetNavContext();
  mEditorContext = mHost.GetEditorContext();
}
```

So far we haven't done much other than ensure that the plugin will behave properly in the Flare UI. Later we'll add another line to actually do something useful.

GetAuthor(), GetDescription(), GetName(), GetVersion()

As you may have already figured out, these four methods will be called to show the author, description, name, and version in the Flare user interface for managing plugins. Each of these should return a string.

```
public string GetAuthor()
{
  return "Flare for Programmers";
}

public string GetDescription()
{
  return "This is a first plugin.";
}

public string GetName()
{
  return "Plugin #1";
}

public string GetVersion()
{
  return "0.1";
}
```

Plugins are managed from Flare through **File → Options → Plugins** in the Ribbon UI and **Tools → Options → Plugins** in the classic menu UI. A grid lists the Name, Description, and Version for each deployed plugin and gives users the option to activate and deactivate the plugins.

IsActivated()

This method returns the activation status of the plugin. In our example, we will return the value `mActivated`.

```
public bool IsActivated
{
  get { return mActivated; }
}
```

Stop()

This method is called when a user deactivates the plugin. Notice that `Dispose()` is called for `mHost` to clean up.

```
public void Stop()
{
  MessageBox.Show(GetName() + " deactivated.");
  mHost.Dispose();
  mActivated = false;
}
```

Building and deploying a plugin

Believe it or not, our code now satisfies the requirements for a plugin. It doesn't do much, but let's go ahead and run through the exercise of building and deploying it to get an idea of what's involved in those processes. Right-click the project and select **Build**. You should see the following line as the last line in the Output window:

```
========== Build: 1 succeeded, 0 failed, 0 up-to-date, 0 skipped ==========
```

Right-click the project and select **Open Folder** in File Explorer. Then drill into the `bin\Debug` folder and copy `FirstPlugin.dll`. This is your first plugin assembly.

Navigate to the following folder:

`C:\Program Files (x86)\MadCap Software\MadCap Flare V10\Flare.app\Plugins`

This is where you put your plugin DLL. Create a folder with the same name as the DLL (`FirstPlugin`) and place the DLL inside that new folder. Congratulations, your first plugin is now deployed.

Now, open MadCap Flare and navigate to the plugin manager. This is located at **File → Options → Plugins** in the Ribbon UI and **Tools → Options → Plugins** in the classic menu UI. You should see your plugin in the left-hand panel (see Figure 10.8).

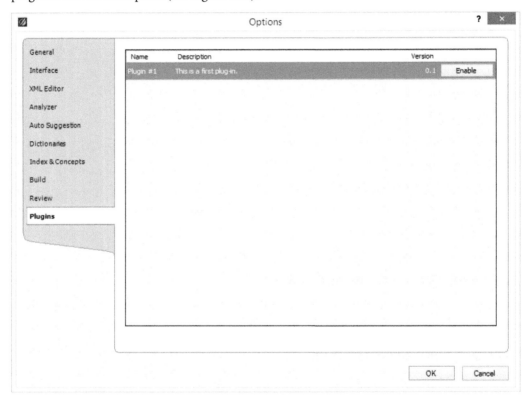

Figure 10.8 – Our first plugin appears in Flare

When you click **Enable**, you get a notice at the bottom of the window.

Note: MadCap recommends that you restart the application to ensure all changes take effect. You don't need to restart for Flare to see the plugin, but if you don't restart, you may get unexpected behavior.

When you click **Disable**, the message defined in `Stop()` will appear (see Figure 10.9).

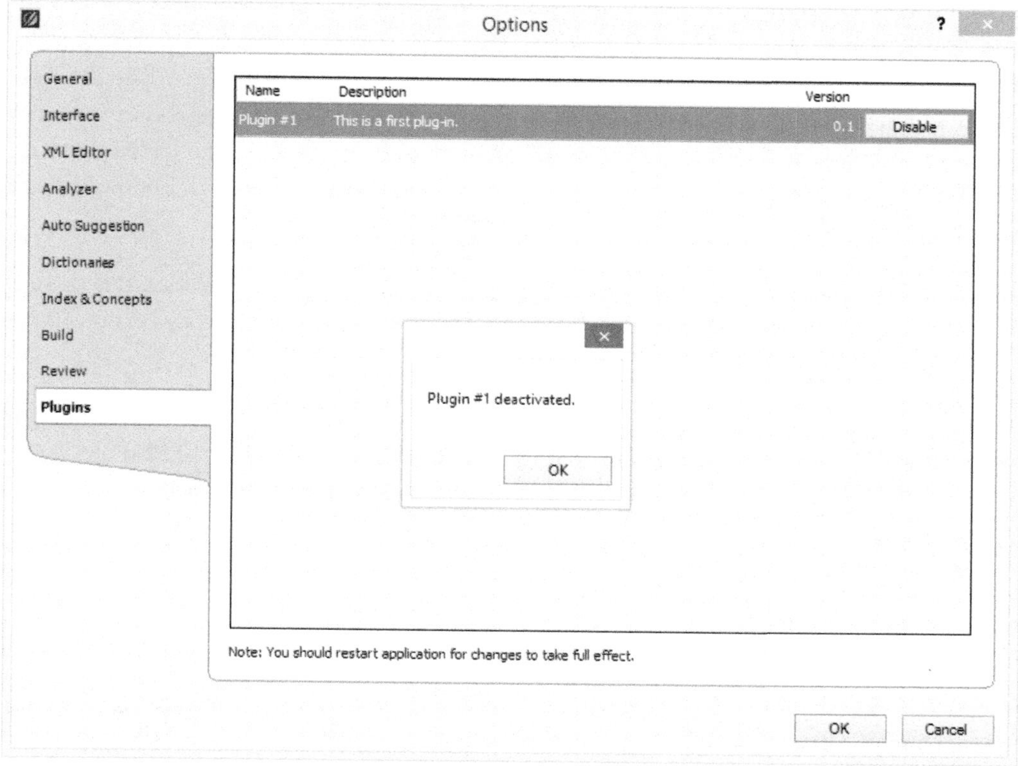

Figure 10.9 – Our first plugin is deactivated

Menus

In this section, we create a toolbar and a ribbon group on a new ribbon, each with a button. This gives us a custom button in either UI.

First, however, you should be aware that there was a compatibility-breaking change between version 9 and version 10 of the API. In particular `ICustomToolstrip` was removed and `ICustomToolBar` was added. This change breaks version 9 plugins that use `ICustomToolstrip`. The good news is that this change enables better code sharing between toolbar buttons and ribbon group buttons.

Before version 10, there was `AddToolStripButton(string, string, EventHandler)`. Now, with version 10, there is `AddButton`, which can use the same class that inherits `ICommand` as `AddRibbonButton(string, ICommand, object, string, RibbonIconSize, string, string, string)`. This change eliminates the need to use the EventHandler and follow a somewhat convoluted path to share logic for the buttons, which was necessary in version 9. Sharing that code is easier with the new version of the API, but dealing with toolbar buttons and buttons on ribbon groups is still a fact of life if you want to support both UI options in Flare, even if some of the code is shared.

Despite the need to recode older plugins that support the Tool Strip user interface, this change is definitely a net positive in terms of code health. We assume that you are using version 10, but we also provide the code for version 9.

Let's look at how to create a toolbar and a ribbon group on a new ribbon, each with a button that, when clicked by a user, will display the message "The button works!" Figure 10.10 shows the result of clicking the button, and Example 10.4 shows a code snippet for this action.

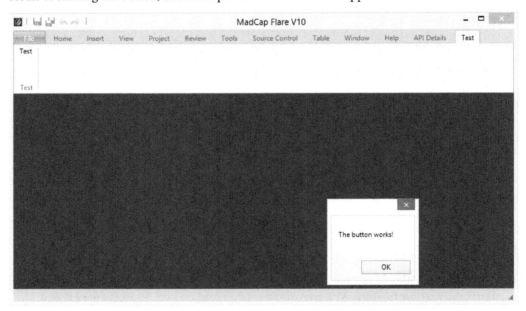

Figure 10.10 – Our second plugin

Example 10.4 – C# Code for creating a toolbar, ribbon, and button command (version 10)

```csharp
private void CreateToolBarAndRibbon()
{
    //ToolBar
    ICustomToolBar toolbar = mNav.CreateCustomToolBar("Test");
    toolbar.AddButton("Test", new ButtonCommand(mEditorContext));

    //Ribbon
    IRibbon ribbon = mNav.GetRibbon();
    IRibbonTab tab = ribbon.AddNewRibbonTab("Test");
    IRibbonGroup group = tab.AddNewRibbonGroup("Test");
    group.AddRibbonButton("Test", new ButtonCommand(mEditorContext),
                null, null, RibbonIconSize.Collapsed, null, null);
}

private class ButtonCommand : ICommand
{
    private IEditorContext passEditorContext;

    public ButtonCommand(IEditorContext pEditorContext)
    {
        passEditorContext = pEditorContext;
    }

    public bool CanExecute(object parameter)
    {
        return true;
    }

    public event EventHandler CanExecuteChanged;

    public void Execute(object parameter)
    {
        MessageBox.Show("The button works!");
    }
}
```

These methods can be placed after the interface implementation in your plugin. One method, CreateToolBarAndRibbon(), creates the toolbar and ribbon; it adds a button to the toolbar for the classic menu UI and a ribbon group that includes a button to the ribbon for the Ribbon UI. Both buttons use the same button command, ButtonCommand, which is the second method. Note that we are grabbing editor context in this command. We're not using it in this example yet, but it will come in handy later.

Now that we've added the code from Example 10.4, we'll update `Execute()` to call the `CreateToolBarAndRibbon` method to create the toolbar and ribbon (see Example 10.5).

Example 10.5 – Modify the `Execute()` function to create the toolbar and ribbon

```
void IPlugin.Execute()
{
  mActivated = true;
  mNav = mHost.GetNavContext();
  mEditorContext = mHost.GetEditorContext();
  CreateToolBarAndRibbon();
}
```

For version 9, we can leave `Execute()` as is. However, to keep terminology consistent, you might want to rename `CreateToolBarAndRibbon()` to be `CreateToolstripAndRibbon()`.

Whatever you call it, this method does need to be modified. The code that creates the ribbon can remain the same, but the code that creates the toolstrip needs to be adjusted for version 9. This includes adding a new method, `button_click_test()`, which is used by the toolstrip. Example 10.6 shows the methods modified for version 9.

In version 9, you could create buttons and combo boxes. In version 10, you can also create menu buttons, separators, split menu buttons, and toggle buttons. Please note the methods for tool strip menu items also changed in version 10. So if your version 9 plugin uses those, you will have to update your code to use the plugin with version 10.

Example 10.6 – C# Code for creating a toolbar, ribbon, and button command (version 9)

```
private void CreateToolBarAndRibbon()
{
    //Toolstrip
    ICustomToolstrip toolstrip = mNav.CreateCustomToolstrip("Test");
    toolstrip.AddToolStripButton("Test", null, button_click_test);

    //Ribbon
    IRibbon ribbon = mNav.GetRibbon();
    IRibbonTab tab = ribbon.AddNewRibbonTab("Test");
    IRibbonGroup group = tab.AddNewRibbonGroup("Test");
    group.AddRibbonButton("Test", new ButtonCommand(mEditorContext),
                null, null, RibbonIconSize.Collapsed, null, null);
}
```

```
private void button_click_test(object sender, EventArgs e)
{
    MessageBox.Show("The button works!");
}

private class ButtonCommand : ICommand
{
    private IEditorContext passEditorContext;

    public ButtonCommand(IEditorContext pEditorContext)
    {
        passEditorContext = pEditorContext;
    }

    public bool CanExecute(object parameter)
    {
        return true;
    }

    public event EventHandler CanExecuteChanged;

    public void Execute(object parameter)
    {
        MessageBox.Show("The button works!");
    }
}
```

Editors

Now let's look at a plugin that we developed for version 9 and updated for version 10. This is also a one button plugin. But this one does a little more. When you have a topic open in the XML Editor with text selected, clicking the button for this plugin shows a screen that describes information about the selection. For example, in Figure 10.11, the sentence "Replace this with your own content." has been selected, and the screen generated by the plugin shows information such as the position of the text, the background and font colors, font family, and font size.

Flare Application and Flare Plugins

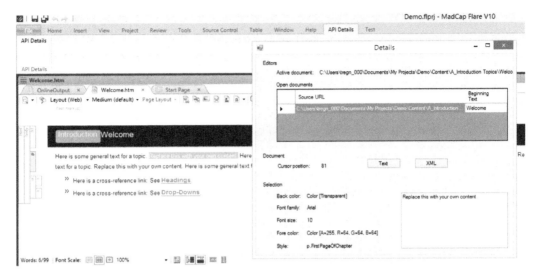

Figure 10.11 – Plugin API details

This plugin shows you any information about the selection that the API has access to. This is useful because most of the available information concerns the text of the selection, and there is no direct access to the DOM for the topic in version 9 of the API. So in version 9, most of what a plugin could perform with the Topic Editor were actions like local formatting or applying style classes.

Note: With Flare version 9, this plugin does not see changes made after it has been invoked. Therefore, if you run this plugin, then make changes using the Topic Editor, you will not see the results of those changes in the plugin output. It is best to avoid making changes with the Topic Editor and the plugin API at the same time.

With version 10, all of that changes. Methods to access and alter the DOM have been added, making version 10 a much stronger API. But before we look at this API, here is Example 10.7, which deals with non-DOM functions.

Example 10.7 – API details code snippet for gathering information returned by the plugin API

```
private static void apiDetailsButtonsActions(IEditorContext pEditorContext)
{
  IDocument activeDocument = pEditorContext.GetActiveDocument();

  ISelection selection;
  String sourceUrl = "";
  String backColor = "";
  String cursorPosition = "";
  String fontfamily = "";
  String fontSize = "";
  String foreColor = "";
  String style = "";
  List<KeyValuePair<string, string>> documents =
    new List<KeyValuePair<string, string>>();
  String documentText = "";
  String documentXml = "";
  string selectionText = "";

  if (activeDocument != null)
  {
    selection = activeDocument.Selection;
    sourceUrl = activeDocument.GetSourceUrl();
    backColor = selection.GetBackColor().ToString();
    cursorPosition = activeDocument.GetCursorPosition().ToString();
    fontfamily = selection.GetFontFamily();
    fontSize = selection.GetFontSize().ToString();
    foreColor = selection.GetForeColor().ToString();
    style = selection.GetStyle();
    documentText = activeDocument.GetDocumentText();
    documentXml = activeDocument.GetDocumentXml();
    selectionText = selection.GetText();

    foreach (IDocument doc in pEditorContext.GetDocuments())
    {
      //will fail if the document doesn't have 10 characters
      documents.Add(new KeyValuePair<string, string>(doc.GetSourceUrl(),
        doc.GetDocumentText().Substring(0, 10)));
    }
  }

  Details DetailsScreen =
    new Details(sourceUrl, backColor, cursorPosition,
        fontfamily, fontSize, foreColor, style,
        documents, documentText, documentXml,
        selectionText);
  DetailsScreen.Show();
}
```

The `apiDetailsButtonsActions()` function, shown in Example 10.7, is called in place of the two `MessageBox.Show()` function calls in Example 10.6. It displays the Windows Form shown in Figure 10.11. You can download a zip file that contains a complete C# plugin with this functionality from http://xmlpress.net/flare/examples/PluginApiDetails.zip.

Remember how we coded editor context into the button command in the second example but didn't use it? Well, now we will use it. This method uses the editor context to get the active document and then uses the active document to get the selection information. It also uses the editor context to get the list of open documents (topics).

The sequence to get to the selection goes `host → editor → document → selection`.

All of that is helpful, but a Flare user can already see most of that information within the application. What we need is a way to make changes. The plugin API provides a variety of methods for making changes. For example, for selections, you can call API methods to change backcolor, font family, font color, forecolor, and span style. A plugin can also make selection text bold, italicized, or underlined. You can specify that selection text be replaced or the length of the selection changed.

For documents, you can have actions that change the stylesheet, select text, and set cursor positions. This functionality existed in version 9 as well. The API documentation you downloaded earlier describes the available methods.

Documents

With Flare version 10, you can also get the MadCap schema and open documents. The function `OpenDocument()` (see Example 10.8) lets you open a topic by path. This method is overloaded so you can open a topic in either Text or XML view.

Example 10.8 uses the function `OpenFileDialog()` in `System.Windows.Forms` to search for and select the path. The path is then passed to `OpenFileDialog()` to open the topic in Text view. `OpenDocument()` is a part of the `IEditor` interface.

Example 10.8 – Open file dialog with a plugin

```
using System;
using System.Collections.Generic;
using System.Linq;
using System.Xml.Linq;
using System.Text;
using System.Windows.Forms;
using System.Windows.Input;
using B3.PluginAPIKit;
using System.IO;

namespace OpenDocumentPlugin
{
 public class SecondPlugin : IPlugin
 {

  private IHost mHost;
  private bool mActivated;
  private INavContext mNav;
  private IEditorContext mEditorContext;

  void IPlugin.Execute()
  {
   mActivated = true;
   mNav = mHost.GetNavContext();
   mEditorContext = mHost.GetEditorContext();
   CreateToolBarAndRibbon();
  }

  public string GetAuthor()
  {
   return "Flare for Programmers";
  }

  public string GetDescription()
  {
   return "This is a plug-in to demonstrate OpenDocument.";
  }

  public string GetName()
  {
   return "Open Document";
  }

  public string GetVersion()
  {
   return "0.1";
  }
```

```csharp
public void Initialize(IHost host)
{
 mHost = host;
}

public bool IsActivated
{
 get { return mActivated; }
}

public void Stop()
{
 MessageBox.Show(GetName() + " deactivated.");
 mHost.Dispose();
 mActivated = false;
}

private void CreateToolBarAndRibbon()
{
 //ToolBar
 ICustomToolBar toolbar = mNav.CreateCustomToolBar("Open Document");
 toolbar.AddButton("Open", new ButtonCommand(mEditorContext));

 //Ribbon
 IRibbon ribbon = mNav.GetRibbon();
 IRibbonTab tab = ribbon.AddNewRibbonTab("Open Document");
 IRibbonGroup group = tab.AddNewRibbonGroup("Open Document");
 group.AddRibbonButton("Open", new ButtonCommand(mEditorContext),
   null, null, RibbonIconSize.Collapsed, null, null);
}

public static void buttonActions(IEditorContext pEditorContext)
{
 OpenFileDialog openFileDialogTopic = new OpenFileDialog();

 openFileDialogTopic.InitialDirectory = "C:\\";
 openFileDialogTopic.Filter = "htm files (*.htm)|*.htm";
 openFileDialogTopic.FilterIndex = 2;
 openFileDialogTopic.RestoreDirectory = true;

 if (openFileDialogTopic.ShowDialog() == DialogResult.OK)
 {
  pEditorContext.OpenDocument(openFileDialogTopic.FileName,
    EditorView.Text);
 }

}


private class ButtonCommand : ICommand
```

```
    {
      private IEditorContext passEditorContext;

      public ButtonCommand(IEditorContext pEditorContext)
      {
        passEditorContext = pEditorContext;
      }

      public bool CanExecute(object parameter)
      {
        return true;
      }

      public event EventHandler CanExecuteChanged;

      public void Execute(object parameter)
      {
        buttonActions(passEditorContext);
      }
    }
  }
}
```

Insert element example

The most important addition to the API in version 10 was DOM manipulation. The plugin in Example 10.9 manipulates the DOM to insert an element as a child of another element and enter a tag name for the inserted element.

Example 10.9 – Insert element example

```
using System;
using System.Collections.Generic;
using System.Linq;
using System.Xml.Linq;
using System.Text;
using System.Windows.Forms;
using System.Windows.Input;
using B3.PluginAPIKit;
using System.Xml;

namespace InsertElementPlugin
{
  public class InsertElementPlugin : IPlugin
  {
    private IHost mHost;
```

```csharp
    private bool mActivated;
    private INavContext mNav;
    private IEditorContext mEditorContext;

    public bool IsActivated
    {
     get { return mActivated; }
    }

    public string GetVersion()
    {
     return "0.1";
    }

    public string GetAuthor()
    {
     return "Flare for Programmers";
    }

    public string GetDescription()
    {
     return "This is a change element tag plug-in.";
    }

    public string GetName()
    {
     return "Change Element Tag";
    }

    public void Initialize(IHost host)
    {
     mHost = host;
    }

    public void Execute()
    {
     mActivated = true;
     mNav = mHost.GetNavContext();
     mEditorContext = mHost.GetEditorContext();
     CreateToolBarAndRibbon();
    }

    public void Stop()
    {
     MessageBox.Show(GetName() + " deactivated.");
     mHost.Dispose();
     mActivated = false;
    }

    private void CreateToolBarAndRibbon()
```

```
{
 //ToolBar
 ICustomToolBar toolbar = mNav.CreateCustomToolBar("Element Tags");
 toolbar.AddButton("Change", new ButtonCommand(mEditorContext));

 //Ribbon
 IRibbon ribbon = mNav.GetRibbon();
 IRibbonTab tab = ribbon.AddNewRibbonTab("Element Tags");
 IRibbonGroup group = tab.AddNewRibbonGroup("Element Tags");
 group.AddRibbonButton("Change",
  new ButtonCommand(mEditorContext),
  null, null, RibbonIconSize.Collapsed, null, null);
}

public static void buttonActions(IEditorContext pEditorContext)
{
 IDocument activeDocument = pEditorContext.GetActiveDocument();
 ISelection selection;

 //We only want to change in the active document. If there isn't one, do nothing.
 if (activeDocument != null)
 {
  //Get the nodes in the selection and find the last element.
  selection = activeDocument.Selection;
  IList<XmlNode> nodes = selection.GetXmlNodeList();
  XmlNode node;
  if (nodes.Last().NodeType.Equals(System.Xml.XmlNodeType.Element))
  {
   node = nodes.Last();
  }
  else
  {
   if (nodes[0].ParentNode.Equals(null))
   {
    node = nodes.Last();
   }
   else
   {
    node = nodes.Last().ParentNode;
   }
  }

  //Get the new tag name
  string input = Microsoft.VisualBasic.Interaction.InputBox("Enter new tag",
    "Tag", "Default", -1, -1);

  //Create the new element and use ReplaceDocumentNode
  XmlElement el = (XmlElement)node;
  XmlAttributeCollection attrs = el.Attributes;
  string inner = el.InnerXml;
```

```
    XmlElement newEl = activeDocument.GetXmlDocument().CreateElement(input);
    newEl.InnerXml = inner;
    activeDocument.SetCursorPosition(0);
    activeDocument.Selection.SetSelectionLength(0);
    activeDocument.ReplaceDocumentNode(node, newEl, true);
    //activeDocument.InsertDocumentNode(node,
    // newEl, activeDocument.GetCursorPosition(), true);
    activeDocument.UpdateView();
  }
 }
 private class ButtonCommand : ICommand
 {
  private IEditorContext passEditorContext;

  public ButtonCommand(IEditorContext pEditorContext)
  {
   passEditorContext = pEditorContext;
  }

  public bool CanExecute(object parameter)
  {
   return true;
  }

  public event EventHandler CanExecuteChanged;

  public void Execute(object parameter)
  {
   buttonActions(passEditorContext);
  }
 }
}
```

Overall, these examples should help show you the power of Flare plugins and the Flare API. The possibilities are endless.

CHAPTER 11
Formatting and Pasting Code Samples into Flare

This book is mostly about programming, scripting, and automating in the context of MadCap Flare. But if you write code, you need to document it. Sometimes manually pasting a code sample inline with an explanation is sufficient. However, regardless of how you create documentation, your code samples need consistent indentation and nicely formatted whitespace to be easily read and understood. Here are some thoughts on how to make code samples that work well for reader and a plugin for pasting code samples.

Use a monospaced font. Proportional fonts, such as Times New Roman, do not work well with code because programmers usually format their code with spaces to align items such as parameters or attributes vertically. Proportional fonts break those alignments. Courier New, Monaco, Consolas, and Inconsolata are common font choices for code.

Take advantage of your editing tool. Many integrated development environments (IDEs) provide auto-formatting. For example, with code open in an editor in Visual Studio, you can select **Edit → Advanced → Format Document** (**Ctrl-K, Ctrl+D**). In NetBeans, you can select **Source → Format** (**Alt+Shift+F**). These actions will format your code according to a set of rules based on the programming language. Other editing tools and IDEs offer similar capabilities.

Plugins for editors such as Notepad++ are available to format various languages. For example, Visual Studio does not format Transact-SQL out-of-the-box, but a formatter for Transact-SQL, Poor Man's T-SQL Formatter, is available for Notepad++ and SQL Server Management Studio as a plugin and an add-in respectively. There are also service websites that will format and add syntax coloring to your code samples.

Some editors will colorize your code to highlight syntax, and sometimes, syntax coloring will be preserved when you copy/paste from an IDE editor to an authoring tool. For example, code pasted from Visual Studio to Microsoft Word will retain coloring. However, copy/paste to Flare does not preserve coloring. In general, don't expect formatting other than whitespace to be preserved in a copy/paste operation between editors. And since syntax coloring is an embellishment applied by the editor automatically, it has no impact on how code is compiled or executed and is not stored in the source file.

Although a code sample can be colorized for syntax in the source, a colorizer that executes at run-time saves authoring time. This may not be possible for all outputs. But for web-based outputs, JavaScript solutions work well. For example, Google offers a library called `prettify`. Instructions are available at MadBlog[1] that show how to use Google `prettify` with Flare HTML5 output.

Whitespace, however, is another matter entirely. Whitespace is defined by the programmer and stored in the source file for most languages. And some languages go so far as to use whitespace to define structure. But for the most part, whitespace is simply an aid to readability, and programmers have a great deal of latitude in how they use it.

Variations in case can also convey useful information. For example, some programmers begin local variables in lower case and global variables in upper case. All of the languages used in this book are case sensitive, but there are languages where some or all of the syntax is case insensitive. In those languages, you need to be careful that you do not create variables or other named entities where the only difference is case. While the case may not affect the execution or compilation of the program, you should pause before making changes. Was your choice of case arbitrary, dictated by another standard, or chosen for a specific purpose?

Sometimes formatters are configurable. For some situations, code should break at 40 or 80 characters per line. Perhaps keywords should always be capitalized. These types of conventions depend on the programming language and the audience.

You may need to break a long line, especially if you are creating content for print. Some languages are more forgiving about line breaks than others. For example, many languages, such as C and C++, treat newline characters the same as spaces, unless they are inside a quoted string. Python uses a backslash at the end of a line to indicate a continuation, and Visual Basic uses a space and an underscore. It is often helpful to break up a line between parameters or operators.

Now, supposing the code sample is satisfactorily formatted, how do you insert it into Flare? Here's one approach:

1. **Copy**
2. **Paste → Inline text**
3. Change the tag to `<pre>`

---

[1] http://www.madcapsoftware.com/blog/2012/08/14/syntax-highlighting-using-prettify-a-syntax-highlighter-by-google/

Pretty simple. Pasting as inline text places everything in a single element. The `<pre>` element is the HTML tag for pre-formatted text. With no other styles applied, characters in pre-formatted text, including whitespace, are usually rendered by browsers as is. More recently, the HTML5 committees have recommended the `<code>` tag in conjunction with the `<pre>` tag.

How about doing that with a single click of a button? Example 11.1 is a Flare plugin that takes text from clipboard, wraps it in a `<pre>` element and pastes it into a Flare topic.

Plugin for pasting code samples

Example 11.1 – Sample C# plugin for pasting code samples

```
using System;
using System.Collections.Generic;
using System.Linq;
using System.Xml.Linq;
using System.Text;
using System.Windows.Forms;
using System.Windows.Input;
using B3.PluginAPIKit;
using System.Xml;

namespace CodePastePlugin
{
 public class CodePastePlugin : IPlugin
 {
  private IHost mHost;
  private bool mActivated;
  private INavContext mNav;
  private IEditorContext mEditorContext;

  public bool IsActivated
  {
   get { return mActivated; }
  }

  public string GetVersion()
  {
   return "0.1";
  }

  public string GetAuthor()
  {
   return "Flare for Programmers";
  }
```

```
public string GetDescription()
{
 return "This is a code paste plug-in.";
}
public string GetName()
{
 return "Code Paste";
}
public void Initialize(IHost host)
{
 mHost = host;
}
public void Execute()
{
 mActivated = true;
 mNav = mHost.GetNavContext();
 mEditorContext = mHost.GetEditorContext();
 CreateToolBarAndRibbon();
}
public void Stop()
{
 MessageBox.Show(GetName() + " deactivated.");
 mHost.Dispose();
 mActivated = false;
}
private void CreateToolBarAndRibbon()
{
 //ToolBar
 ICustomToolBar toolbar = mNav.CreateCustomToolBar("Code Paste");
 toolbar.AddButton("Paste", new ButtonCommand(mEditorContext));

 //Ribbon
 IRibbon ribbon = mNav.GetRibbon();
 IRibbonTab tab = ribbon.AddNewRibbonTab("Code Paste");
 IRibbonGroup group = tab.AddNewRibbonGroup("Code Paste");
 group.AddRibbonButton("Paste", new ButtonCommand(mEditorContext),
  null, null, RibbonIconSize.Collapsed, null, null);
}
public static void buttonActions(IEditorContext pEditorContext)
{
 IDocument activeDocument = pEditorContext.GetActiveDocument();
 ISelection selection;
```

```csharp
//We only want to insert into the active document.
//If there isn't one, do nothing.
if (activeDocument != null)
{
 //Get the nodes in the selection and find the last element.
 selection = activeDocument.Selection;
 IList<XmlNode> nodes = selection.GetXmlNodeList();
 XmlNode parentNode;
 if (nodes.Last().NodeType.Equals(System.Xml.XmlNodeType.Element))
 {
  parentNode = nodes.Last();
 }
 else
 {
  if (nodes[0].ParentNode.Equals(null))
  {
   parentNode = nodes.Last();
  }
  else
  {
   parentNode = nodes.Last().ParentNode;
  }
 }
 //Create a code element that contains copied text.
 XmlElement el =
  activeDocument.GetXmlDocument().CreateElement("code");
 el.InnerText =
  Clipboard.GetText(System.Windows.Forms.TextDataFormat.UnicodeText);
 //Insert the code element
 //at the end of the last element and update the view.
 activeDocument.InsertDocumentNode(parentNode, (XmlNode)el,
  activeDocument.GetCursorPosition(), false);
 activeDocument.UpdateView();
 }
}


private class ButtonCommand : ICommand
{
 private IEditorContext passEditorContext;

 public ButtonCommand(IEditorContext pEditorContext)
 {
  passEditorContext = pEditorContext;
 }

 public bool CanExecute(object parameter)
 {
  return true;
 }
```

```
    public event EventHandler CanExecuteChanged;

    public void Execute(object parameter)
     {
      buttonActions(passEditorContext);
     }
    }
   }
  }
 }
```

CHAPTER 12
Strategies for Combining Generated and Authored Content

Much of this book describes possible points of intervention while authoring or generating content. But what general approach should you use? You may see an example that describes how to add snippets to authored content programmatically or read a description of how to generate Flare topics that contain content pulled from a database. But how do you integrate these techniques with your authoring processes? This chapter describes some possible strategies.

For the purposes of this chapter, *authored* means something typed into the Topic Editor in Flare by a human being; *generated* means a Flare artifact, such as a topic, created by program or a script; and *refreshed* means generated content that has been regenerated and copied back into the same Flare project, overwriting the previous version.

Strategy: Don't touch generated content

With this approach, we assume that authored content is always authored and generated content is never authored. You only deal with generated content when you start a build or copy generated content into a project. But you never edit generated topics or generated TOCs. With this approach, there is no ambiguity about what an author can change. You never tweak a generated reference, and you never link to or include authored content in generated content. However, you can link to or include generated content in authored content.

Consider a workflow with an automated process that creates Flare reference topics. For example, you might have a database that contains information about cities, such as population demographics and business locations. A Java application could connect to that database, retrieve data about each city, and generate a Flare topic for each city that includes the population and business information. Those topics could be arranged alphabetically in a generated TOC.

You could place those topics and the TOC in a Flare project and generate a Flare help system. But what if you want to add some authored content, such as sibling topics that describe each city from a human perspective to complement the statistical content?

Using this strategy, you would leave the generated TOC in place and author a new one. You would only edit the manually created TOC and never alter generated topics or the generated TOC. However, you could open the generated TOC and drag entries from the generated TOC into the manually created TOC. Adding a generated topic to the manually created TOC does not alter the generated TOC or the generated topic. But you have integrated generated content with authored content, which is a powerful capability.

And you don't have to stop at mixing and matching generated and authored topics. If you want, you can include the entire generated TOC in the authored TOC without altering the generated TOC. When you build a help system or print output from the authored TOC, you end up with well-maintained data from a data source integrated with human content.

Figure 12.1 shows both generated and authored content, with the authored content linking to (or including) generated content. The reverse operation, linking from generated content to authored content, is shown as not allowed.

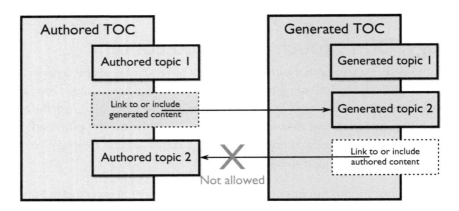

Figure 12.1 – Illustrating the strategy: Don't touch generated content

Variations

With this strategy, you normally build a single project that contains everything you need, both authored and generated content. You can vary the strategy by having separate projects for generated and authored content, then importing generated topics into the authored content or linking from the authored project to the generated project. We don't recommend either of these variations.

Strategy: Generate once and edit freely

With this strategy, content is not refreshed. You generate content once, then use it as the basis for all future updates (see Figure 12.2). The idea is that, for some projects, generating content once and then carrying on from there can be more cost effective than maintaining a workflow that segregates authored and generated content and then merges future changes to authored content with generated content. Of course, in this case, all future updates to the content will have to rely on human intervention. You can still generate new content, but you need to include that content manually.

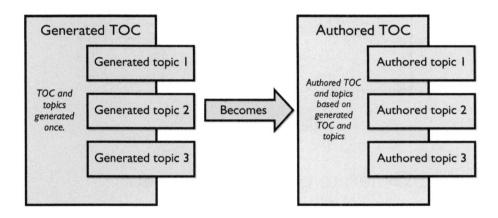

Figure 12.2 – Illustrating the strategy: Generate once and edit freely

The biggest advantage of this strategy is that you have a great deal of flexibility in generating and organizing content. You can make as many changes as you want without any concern about interfering with future processing because there won't be any future processing.

The biggest disadvantage of this strategy is that you can no longer rely on the system for updates. Any updates require human intervention, even if the update is just running a content generator and bringing that content into the project.

Strategy: Generate content but only keep the topics

With this strategy, you generate topics and manually organize them in an authored TOC. (see Figure 12.3). Since the TOC is authored, you must add and remove topics manually. You refresh topics by overwriting the existing version, and you can include authored content alongside the generated content.

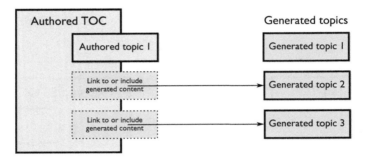

Figure 12.3 – Illustrating the strategy: Generate content but only keep the topics

Strategy: Generate content as snippets

With this approach, you generate content as snippets, rather than topics. Snippets can be included in authored topics and other snippets (see Figure 12.4). This is similar to the strategy described in the section titled "Strategy: Generate once and edit freely," but the level of granularity is different.

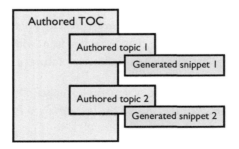

Figure 12.4 – Illustrating the strategy: Generate content as snippets

Strategy: Generate topics and snippets

With this approach, you generate at least three artifacts for each topic (see Figure 12.5). You generate a wrapper topic and at least two snippets. One of the snippets is intended for authored content. You generate that snippet once, then edit it manually going forward. The other snippet is meant to be refreshed, which means you do not manually author it, and you may end up generating it multiple time. This strategy lets you keep authored content separate from generated content, but have both combined in the Flare project and your output.

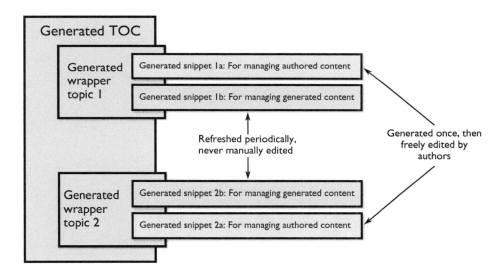

Figure 12.5 – Illustrating the strategy: Generate topics and snippets

Conclusion

There are many other potential variations and strategies than we have discussed here. But these strategies handle most of the situations we have encountered and provide some clear patterns you can apply to your content.

Acknowledgments

We would like to thank everyone at MadCap Software. Jose Sermeno graciously answered many questions and put us in touch with the right people. Rob Hollinger and his team included us in the betas for recent versions of Flare. Francis Novak encouraged the Flare for Programmers blog in its beginning and gave Tom a chance to visit MadCap during the first MadWorld conference.

Thanks to Anthony Olivier for contributing the Foreword and for leading the team at MadCap. Thanks to our technical reviewers – Nita Beck, Jose Sermeno, and Alan Williamson – who gave their time to read and comment on a draft of this book and to our editor, Richard Hamilton. We would also like to thank the MadCap user community, which actively participates on the forums, comments at MadBlog, and participates on the LinkedIn group. And thank you to everyone who has read and commented on the Flare for Programmers blog.[1]

Thomas Tregner

Thank you to my wife Laura for patiently supporting me while working out examples and writing after the kids went to sleep.

David Owens

I would like to thank my wife, Theresa, and our two sons, Jake and Sam, for giving up family time for me to work on this project. Thanks to Thomas for inviting me to collaborate on the book. Also thanks to Mary Beth Westmoreland for her continued support and to the Blackbaud User Education team for giving me the opportunity to work with some of the best professionals in the business on a daily basis.

---

[1] http://tregner.com/flare-blog/

APPENDIX A
Batch File Generator Utility

The batch file commands manager utility, which is discussed in Chapter 8, *Document Automation and Batch Files* (p. 83), collects information from the user, then starts at the root folder and recursively finds Flare target files (*.fltar). It builds batch commands for each target using the user-supplied information.

Example A.1 shows the code for the utility. The code can be downloaded from the companion website (http://xmlpress.net/flare/examples).

Example A.1 – C# code for the Flare batch commands manager

```
Imports System.IO
Imports System.Xml
Imports System.Xml.Serialization

Public Class FormFlareBatchCommandsManager

Private Sub ButtonSelectRootFolder_Click(ByVal sender As System.Object,
                                    ByVal e As System.EventArgs) _
                            Handles ButtonSelectRootFolder.Click
    'Populate a label for the root folder with a selection from a file dialog.
    'The root folder is where the targets will be found.
    'Includes subfolders.
    If FolderBrowserDialogRootFolder.ShowDialog() = DialogResult.OK Then
      LabelRootFolderPath.Text = FolderBrowserDialogRootFolder.SelectedPath
      RefreshGrid(LabelRootFolderPath.Text)
    End If
End Sub

Private Sub ButtonFlareAppPath_Click(ByVal sender As System.Object,
                                    ByVal e As System.EventArgs) _
                            Handles ButtonFlareAppPath.Click
    'Populate a label for the path to Flare.app.
    'This is how the manager determines the location of madbuild.exe.
    If FolderBrowserDialogFlareAppPath.ShowDialog() = DialogResult.OK Then
      LabelFlareAppPath.Text = FolderBrowserDialogFlareAppPath.SelectedPath
    End If
End Sub

Private Sub ButtonBATFilesFolder_Click(ByVal sender As System.Object,
                                    ByVal e As System.EventArgs) _
                            Handles ButtonBATFilesFolder.Click
    'Populate a label for the path to the folder for the generated BAT files.
```

```vb
      If FolderBrowserDialogBATFilesFolder.ShowDialog() = DialogResult.OK Then
        LabelBATFilesFolder.Text = FolderBrowserDialogBATFilesFolder.SelectedPath
      End If
    End Sub

    Private Sub ButtonTFPath_Click(ByVal sender As System.Object,
                                   ByVal e As System.EventArgs) _
                       Handles ButtonTFPath.Click
      'Populate a label for the path to the folder that contains TF.exe,
      'the command executable for Team Foundation Server.
      'This is to support adding get commands for TFS.
      If FolderBrowserDialogTFPath.ShowDialog() = DialogResult.OK Then
        LabelTFPath.Text = FolderBrowserDialogTFPath.SelectedPath
      End If
    End Sub

    Private Sub RefreshGrid(ByVal rootPath As String)
      'Clear the grid and populate it with all of the targets in the folder.
      DataGridViewTargets.Rows.Clear()
      For Each foundFile _
        As String In My.Computer.FileSystem.GetFiles(rootPath,
          Microsoft.VisualBasic.FileIO.SearchOption.SearchAllSubDirectories,
          "*.fltar")
        'This if excludes the prefix we use for template targets.
        If Not Path.GetFileName(foundFile).Contains("ztemplate") Then
          Dim directoryName = Path.GetDirectoryName(foundFile)
          'finding the Targets folder part in the path.
          Dim projTarg As String
          projTarg = "Project\Targets"
          Dim indexForProjTarg = directoryName.IndexOf(projTarg)
          directoryName = directoryName.Substring(0, indexForProjTarg)
          Dim targetNameWithFolder As String
          targetNameWithFolder = _
            Path.GetDirectoryName(foundFile).Substring(indexForProjTarg + _
              15).Replace("\", "/") _
              + "/" + Path.GetFileName(foundFile)
          If targetNameWithFolder.Substring(0, 1) = "/" Then
            targetNameWithFolder = targetNameWithFolder.Substring(1)
          End If
          Dim projectFileName = ""
          'Find the project file for the target.
          'Depending on how flexible Flare gets with the file locations,
          'this may need to be revisited.
          For Each foundProjectFile _
            As String In My.Computer.FileSystem.GetFiles(directoryName,
              Microsoft.VisualBasic.FileIO.SearchOption.SearchTopLevelOnly,
              "*.flprj")
            projectFileName = foundProjectFile
          Next
          'Add a row to the grid for the target.
```

```vb
            DataGridViewTargets.Rows.Add(True, Path.GetFileName(foundFile), _
                                         projectFileName, _
                                         ReturnType(foundFile).ToString, _
                                         targetNameWithFolder)
      End If
   Next
End Sub

Private Sub FormFlareBatchCommandsManager_Load(ByVal sender As System.Object, _
                                               ByVal e As System.EventArgs) _
                                    Handles MyBase.Load
   'Populate the grid on load of the form.
   RefreshGrid(My.Settings.RootFolderPath)
End Sub

Private Sub ButtonGetNow_Click(ByVal sender As System.Object, _
                                ByVal e As System.EventArgs) _
                      Handles ButtonGetNow.Click
   'Get latest using TF.exe. Gets for the root folder on the form.
   Dim getLines As String
   getLines = vbCrLf + "cd " + LabelTFPath.Text.ToString + vbCrLf
   getLines = getLines + "TF.exe get " + LabelRootFolderPath.Text.ToString + _
              " /recursive" + vbCrLf
   getLines = getLines + "Pause"
   Dim file As System.IO.FileStream
   Dim batchPath = LabelBATFilesFolder.Text.ToString + "\" + "TFGet.bat"
   file = System.IO.File.Create(batchPath)
   file.Close()
   My.Computer.FileSystem.WriteAllText(batchPath, getLines, True, _
                                       New System.Text.UTF8Encoding(True))
   Shell(batchPath)
End Sub

Private Function getOutputFolder(ByVal targetFile As String)
   'Return the attribute value for OutputFolder in the CatapultTarget
   'element (root) or if there is not one, return the text in the label
   'for the folder to which to output the batch files
   Dim targetOutputFolder As String
   Dim flareTargetXHTML As XDocument = XDocument.Load(targetFile)
   If Not flareTargetXHTML.Root.Attribute("OutputFolder") Is Nothing Then
      targetOutputFolder = flareTargetXHTML.Root.Attribute("OutputFolder").Value
      Return targetOutputFolder
   Else
      Return LabelBATFilesFolder.Text
   End If
End Function

Private Sub CreateMultipleBatchFiles()
   'This creates many files that each build one target.
```

```vb
'Generate a batch file for every target in the grid rows.
For Each row As DataGridViewRow In DataGridViewTargets.Rows
  If Not row.Cells(1).Value = "" Then
    'Create a BAT file using the project and target name.
    Dim file As System.IO.FileStream
    Dim batchPath As String
    Dim fileNameFragment As String
    fileNameFragment = row.Cells(1).FormattedValue.ToString
    Dim targetFileName = fileNameFragment.Substring(0, _
                    fileNameFragment.Length - 6)
    fileNameFragment = "_" + _
      fileNameFragment.Substring(0, fileNameFragment.Length - 6)

    Dim fileNameFragment2 As String
    fileNameFragment2 = Path.GetFileName(row.Cells(2).FormattedValue.ToString)
    fileNameFragment2 = "_" + _
      fileNameFragment2.Substring(0, fileNameFragment2.Length - 6)
    fileNameFragment = fileNameFragment2 + fileNameFragment
    batchPath = LabelBATFilesFolder.Text.ToString + "\" + _
        "FlareTargetBatch" + fileNameFragment + ".bat"
    file = System.IO.File.Create(batchPath)
    file.Close()
    Dim comment As String
    comment = "REM comment goes here" + vbCrLf
    My.Computer.FileSystem.WriteAllText(batchPath, _
                    comment, True, New System.Text.UTF8Encoding(False))

    'Add a line to the batch file to change the directory to the TF.exe path
    'and a line to run TF.exe get on the root folder.
    If CheckBoxGetLatestVersion.Checked Then
      Dim cdString As String
      cdString = "cd " + LabelTFPath.Text.ToString + vbCrLf
      My.Computer.FileSystem.WriteAllText(batchPath, cdString, True)
      Dim getPath As String
      getPath = "TF.exe get " + LabelRootFolderPath.Text.ToString + _
        " /recursive" + vbCrLf
      My.Computer.FileSystem.WriteAllText(batchPath, getPath, True)
    End If

    'Add a line to the batch file to change the directory to the
    'madbuild.exe path and a command line for the target in the row.
    'Include -log true if selected.
    Dim flareAppPath As String
    flareAppPath = "cd " + LabelFlareAppPath.Text.ToString + vbCrLf
    My.Computer.FileSystem.WriteAllText(batchPath, flareAppPath, True)

    Dim buildLine As String
    buildLine = "madbuild -project " + row.Cells(2).FormattedValue.ToString + _
            " -target " + row.Cells(4).FormattedValue.ToString
```

```vb
      If CheckBoxAddLogTrue.Checked = True Then
        buildLine = buildLine + " -log true"
      End If

      buildLine = buildLine + vbCrLf

      If row.Cells(0).Value = True Then
        My.Computer.FileSystem.WriteAllText(batchPath, buildLine, True)
      Else
        buildLine = "rem " + buildLine
        My.Computer.FileSystem.WriteAllText(batchPath, buildLine, True)
      End If

      'Add lines to place the output in a ZIP file.
      If CheckBoxCreateZIP.Checked Then
        My.Computer.FileSystem.WriteAllText(batchPath, "cd " + _
          Label7ZIPPath.Text + vbCrLf, True)
        Dim folderToArchive As String
        Dim pathOfTarget As String
        pathOfTarget = row.Cells(4).FormattedValue.ToString
        folderToArchive = getOutputFolder(pathOfTarget)
        Dim ZIPCommand As String
        ZIPCommand = _
          "REM z target is ZIP filename and source is the path after it" + _
          vbCrLf + "7z a C:\FlareOutput\zip\" + fileNameFragment + _
          ".zip " + folderToArchive + "\" + _
          "Output\" + targetFileName + "\*" + vbCrLf
        My.Computer.FileSystem.WriteAllText(batchPath, ZIPCommand, True)
      End If

      If CheckBoxAddPause.Checked Then
        My.Computer.FileSystem.WriteAllText(batchPath, "Pause", True)
      End If
    End If
  Next

End Sub

Private Sub CreateSingleBatchFile(ByVal fileNameFragment As String)
  'This creates one file that builds many targets.
  'Create a BAT file in the folder indicated on the form.
  'The name is FlareTargetBatch + the fragment passed to this procedure.
  Dim file As System.IO.FileStream
  Dim batchPath As String
  batchPath = LabelBATFilesFolder.Text.ToString + _
    "\" + "FlareTargetBatch" + fileNameFragment + ".bat"
  file = System.IO.File.Create(batchPath)
  file.Close()
  Dim comment As String
  comment = "REM comment goes here" + vbCrLf
```

```vb
    My.Computer.FileSystem.WriteAllText(batchPath,
        comment, True, New System.Text.UTF8Encoding(False))

    'Add a line to the batch file to change the directory
    'to the TF.exe path
    'and a line to run TF.exe get on the root folder.
    If CheckBoxGetLatestVersion.Checked Then
      Dim cdString As String
      cdString = "cd " + LabelTFPath.Text.ToString + vbCrLf
      My.Computer.FileSystem.WriteAllText(batchPath, cdString, True)
      Dim getPath As String
      getPath = "TF.exe get " + LabelRootFolderPath.Text.ToString + _
          " /recursive" + vbCrLf
      My.Computer.FileSystem.WriteAllText(batchPath, getPath, True)
    End If

    'Add a line to the batch file to change the directory
    'to the madbuild.exe path.

    Dim flareAppPath As String
    flareAppPath = "cd " + LabelFlareAppPath.Text.ToString + vbCrLf
    My.Computer.FileSystem.WriteAllText(batchPath, flareAppPath, True)

    'Add a command line for every target in the grid.
    'Include -log true if selected.
    For Each row As DataGridViewRow In DataGridViewTargets.Rows
      Dim buildLine As String
      buildLine = "madbuild -project " +
          row.Cells(2).FormattedValue.ToString +
          " -target " + row.Cells(1).FormattedValue.ToString
      If CheckBoxAddLogTrue.Checked = True Then
        buildLine = buildLine + " -log true"
      End If
      buildLine = buildLine + vbCrLf

      If row.Cells(0).Value = True Then
        My.Computer.FileSystem.WriteAllText(batchPath, buildLine, True)
      Else
        buildLine = "rem " + buildLine
        My.Computer.FileSystem.WriteAllText(batchPath, buildLine, True)
      End If
    Next

    If CheckBoxAddPause.Checked Then
      My.Computer.FileSystem.WriteAllText(batchPath, "Pause", True)
    End If
End Sub
```

```vb
Private Sub ButtonCreateSingleBatchFile_Click(ByVal sender As System.Object, _
                                ByVal e As System.EventArgs) _
                                Handles ButtonCreateSingleBatchFile.Click
    CreateSingleBatchFile(System.Guid.NewGuid.ToString)
End Sub

Private Sub ButtonCreateBatchFileForEach_Click(ByVal sender As System.Object, _
                                ByVal e As System.EventArgs) _
                                Handles ButtonCreateBatchFileForEach.Click
    CreateMultipleBatchFiles()
End Sub

Private Function ReturnType(ByVal targetFile As String)
    'Return the attribute value for Type in the CatapultTarget element (root)
    Dim targetType As String
    Dim flareTargetXHTML As XDocument = XDocument.Load(targetFile)
    targetType = flareTargetXHTML.Root.Attribute("Type").Value
    Return targetType
End Function

Private Sub CheckBoxTFS_CheckedChanged(ByVal sender As System.Object, _
                                ByVal e As System.EventArgs) _
                                Handles CheckBoxTFS.CheckedChanged
    'This doesn't provide much value at this point.
    'If there are other TFS options, this will enable or disable those.
    If CheckBoxTFS.Checked = True Then
        CheckBoxGetLatestVersion.Enabled = True
    Else
        CheckBoxGetLatestVersion.Enabled = False
        CheckBoxGetLatestVersion.Checked = False
    End If
End Sub

Private Sub ButtonSaveConfiguration_Click(ByVal sender As System.Object, _
                                ByVal e As System.EventArgs) _
                                Handles ButtonSaveConfiguration.Click
    'Create an XML file to store the manager's configuration.
    Dim configFlareBatchXML As XDocument = _
<?xml version="1.0" encoding="utf-8" standalone="yes"?>
<!--Flare Batch Commands Manager Configuration File-->
<FlareBatchCommandsConfiguration>

</FlareBatchCommandsConfiguration>

    'Add elements to store the settings in the manager and
    'populate the elements with values from the form.
    Dim RootFolder As XElement
    RootFolder = <RootFolder></RootFolder>
    RootFolder.Value = LabelRootFolderPath.Text
    configFlareBatchXML.Root.Add(RootFolder)
```

```vbnet
Dim FlareApp As XElement
FlareApp = <FlareApp></FlareApp>
FlareApp.Value = LabelFlareAppPath.Text
configFlareBatchXML.Root.Add(FlareApp)
Dim BATFiles As XElement
BATFiles = <BATFiles></BATFiles>
BATFiles.Value = LabelBATFilesFolder.Text
configFlareBatchXML.Root.Add(BATFiles)
Dim TFSControlled As XElement
TFSControlled = <TFSControlled></TFSControlled>
TFSControlled.Value = CheckBoxTFS.Checked.ToString
configFlareBatchXML.Root.Add(TFSControlled)
Dim TFSGetLatest As XElement
TFSGetLatest = <TFSGetLatest></TFSGetLatest>
TFSGetLatest.Value = CheckBoxGetLatestVersion.Checked.ToString
configFlareBatchXML.Root.Add(TFSGetLatest)
Dim TFEXEPath As XElement
TFEXEPath = <TFEXEPath></TFEXEPath>
TFEXEPath.Value = LabelTFPath.Text
configFlareBatchXML.Root.Add(TFEXEPath)
Dim AddLogTrue As XElement
AddLogTrue = <AddLogTrue></AddLogTrue>
AddLogTrue.Value = CheckBoxAddLogTrue.Checked.ToString
configFlareBatchXML.Root.Add(AddLogTrue)
Dim AddPause As XElement
AddPause = <AddPause></AddPause>
AddPause.Value = CheckBoxAddPause.Checked.ToString
configFlareBatchXML.Root.Add(AddPause)
Dim CreateZIP As XElement
CreateZIP = <ZIPPath></ZIPPath>
CreateZIP.Value = CheckBoxCreateZIP.Checked.ToString
configFlareBatchXML.Root.Add(CreateZIP)
Dim ZIPPath As XElement
ZIPPath = <ZIPPath></ZIPPath>
ZIPPath.Value = Label7ZIPPath.Text
configFlareBatchXML.Root.Add(ZIPPath)
'This one has children since it represents the grid on the form.
'Only store the Include setting, target,
'and project (first three columns)
'since this is all we need to identify.
Dim Commands As XElement
Commands = <Commands></Commands>
'Loop through each row in the grid and create an element and child elements.
For Each row In DataGridViewTargets.Rows
   Dim newRow As XElement
   Dim includeCell As XElement
   Dim targetCell As XElement
   Dim projectCell As XElement
   newRow = <CommandInfo></CommandInfo>
   includeCell = <Include></Include>
```

```vbnet
        includeCell.Name = "Include"
        includeCell.Value = row.Cells(0).FormattedValue.ToString
        newRow.Add(includeCell)
        targetCell = <Target></Target>
        targetCell.Name = "Target"
        targetCell.Value = row.Cells(4).FormattedValue.ToString
        newRow.Add(targetCell)
        projectCell = <Project></Project>
        projectCell.Name = "Project"
        projectCell.Value = row.Cells(2).FormattedValue.ToString
        newRow.Add(projectCell)
        Commands.Add(newRow)
    Next
    configFlareBatchXML.Root.Add(Commands)
    'Save the configuration file. Limit the dialog to XML.
    SaveFileDialogConfig.FileName = LabelBATFilesFolder.Text + _
                         "\SpecialConfiguration.xml"
    SaveFileDialogConfig.Filter = "XML (*.xml)|*.xml"
    If SaveFileDialogConfig.ShowDialog() = DialogResult.OK Then
        configFlareBatchXML.Save(SaveFileDialogConfig.FileName)
    End If
End Sub

Private Sub ButtonLoadConfiguration_Click(ByVal sender As System.Object, _
                                ByVal e As System.EventArgs) _
                            Handles ButtonLoadConfiguration.Click
    'Open the dialog to select the configuration file.
    OpenFileDialogConfig.Filter = "XML (*.xml)|*.xml"

    'Loop through the elements to populate the form.
    'The grid will be populated by the selection of RootFolder.
    'The settings for Include come from the file.
    If OpenFileDialogConfig.ShowDialog() = DialogResult.OK Then
        Dim configFlareBatchXML As XDocument = _
            XDocument.Load(OpenFileDialogConfig.FileName)
        For Each element In configFlareBatchXML.Root.Elements

            If element.Name = "RootFolder" Then
                LabelRootFolderPath.Text = element.Value
            ElseIf element.Name = "FlareApp" Then
                LabelFlareAppPath.Text = element.Value
            ElseIf element.Name = "BATFiles" Then
                LabelBATFilesFolder.Text = element.Value
            ElseIf element.Name = "TFEXEPath" Then
                LabelTFPath.Text = element.Value
            ElseIf element.Name = "ZIPPath" Then
                Label7ZIPPath.Text = element.Value
            End If

            If element.Name = "TFSControlled" Then
```

```vbnet
            CheckBoxTFS.Checked = element.Value
        End If

        If element.Name = "TFSGetLatest" Then
            CheckBoxGetLatestVersion.Checked = element.Value
        End If

        If element.Name = "AddPause" Then
            CheckBoxAddPause.Checked = element.Value
        End If

        If element.Name = "CreateZIP" Then
            CheckBoxCreateZIP.Checked = element.Value
        End If

        If element.Name = "AddLogTrue" Then
            CheckBoxAddLogTrue.Checked = element.Value
        End If

        If element.Name = "Commands" Then
            For Each element2 In element.Descendants
                For Each row In DataGridViewTargets.Rows
                    If row.Cells(1).FormattedValue.ToString = _
                        element2.Descendants("Target").Value And
                      row.Cells(2).FormattedValue.ToString = _
                        element2.Descendants("Project").Value Then
                        row.Cells(0).Value = element2.Descendants("Include").Value
                    End If
                Next
            Next
        End If
    Next
  End If
End Sub

End Class
```

APPENDIX B
Element List

AutoExcludeNonTaggedFiles

> `AutoExcludeNonTaggedFiles` is an attribute of `<CatapultProjectImport>` in a Flare Project Import File. When set to `true`, Flare will only import files that are tagged with conditions specified in the `ConditionTagExpression` attribute. Setting this attribute is the same as checking the **Auto-include linked files** option in the Project Import Editor.

CatapultAliasFile

> The root element of an alias (`*.flali`) file. A `<CatapultAliasFile>` contains `<Map>` elements with `Name`, `Link`, and `Skin` (optional) attributes. These elements map an identifier to a relative path for a topic. The identifier is mapped to a numerical value in a header file.

CatapultConditionTagSet

> The root element of a Flare Condition Tag Set (`*.flcts`) file. This element contains `<ConditionTag>` sub-elements.

CatapultDestination

> The root element of a Flare Destination (`*.fldes`) file. `<CatapultDestination>` has attributes for the field values in the Flare application's Destination Editor. For example, `Type="ftp"` corresponds to a value of FTP in the **Type** drop down in the Destination Editor.

CatapultProject

> The root element of a Flare Project (`*.flprj`) file. `<CatapultProject>` contains attributes that define basic properties of a Flare project, such as **Project Properties → Master Stylesheet (MasterStylesheet)**, which is the master stylesheet.

CatapultProjectImport

> The root element of a Flare Project Import (`*.flimpfl`) file. `<CatapultProjectImport>` contains attributes that correspond to field values in the Flare application's Project Import Editor. For example, setting `AutoSync="true"` is the same as setting the **Auto-reimport before Generate Output** option in the Project Import Editor. The `<CatapultProjectImport>` element has five child elements: `<Files>`, `<ExcludedFiles>`, `<CollectedSourceFiles>`, `<SkippedSourceFiles>`, and `<GeneratedFiles>`, which contain lists that identify what files to include or exclude in subsequent imports.

CatapultSkin

The root element of a Flare Skin (*.flskn) file. The <CatapultSkin> element contains attributes that correspond to field values in the Flare application's Skin Editor. For example, the SkinType attribute will have a value of WebHelp2 for HTML5 skins (WebHelp 2 was the original name for the HTML5 output type before it was released). A <CatapultToc> can contain other elements, including <Index> and <HtmlHelpOptions>. These elements contain values for additional skin options.

CatapultTarget

The root element of a Flare (*.fltar) file. The <CatapultTarget> element contains attributes that correspond to field values in the Flare application's Target Editor. For example, the Skin attribute corresponds to the **Skin** field in the Target Editor. There may also be a child <PrintedOutput> element for some of those options. Many of the values that default will not appear as attributes until a value is selected or entered.

CatapultToc

The root element of a Flare TOC (*.fltoc) file. A <CatapultToc> contains <TocEntry> elements. The Catapult part of the element name doesn't have any significant meaning. But a valid Flare TOC must have a <CatapultToc> root element.

CatapultVariableSet

The root element of a Flare Variable Set (*.flvar) file. A <CatapultVariableSet> contains <Variable> elements.

ConditionTag

A <ConditionTag> element represents a <Condition> Tag in a Flare Condition Tag Set (*.flvar) file. A <ConditionTag> contains the following attributes: Name, Comment (optional), and BackgroundColor.

ConditionTagExpression

ConditionTagExpression is an attribute of the <CatapultProjectImport> element in a Flare Project Import File. The following line defines an expression to include items with the **Default.ScreenOnly** condition tag set and exclude those with **Default.PrintOnly** set:

ConditionTagExpression="include[Default.ScreenOnly] exclude[Default.PrintOnly] "

This expression is created when **Import Conditions** is edited from the Project Import Editor.

DeleteStale

`DeleteStale` is an attribute of `<CatapultProjectImport>` in a Flare Project Import File. When set to `true`, Flare will delete files that are no longer in the source project on import. This is the same as checking the **Delete stale files** option in the Project Import Editor.

DitaImport

The root element of a DITA Import (`*.flimpdita`) file. The `<DitaImport>` element contains attributes that correspond to the field values in the DITA Import Editor. `<DitaImport>` can contain child elements, such as `<Files>` or `<GeneratedFiles>`, which contain lists that act as bookkeeping mechanisms for the import process.

GlossaryEntry

The `<GlossaryEntry>` element defines a single glossary entry in a Flare Glossary (`*.flglo`) file. `<GlossaryEntry>` contains a `<Terms>` element, which contains one or more `<Term>` elements and a `<Definition>` element. The text of the `<Term>` elements are the terms, and text of the `<Definition>` element is the definition that applies for all of the terms.

IncludeLinkedFiles

`IncludeLinkedFiles` is an attribute of `<CatapultProjectImport>` in a Flare Project Import File. When set to `true`, on import, Flare will import any file that an imported file links to. This is the same as checking the **Auto-include linked files** option in the Project Import Editor.

IncludePattern

`IncludePattern` is an attribute of `<CatapultProjectImport>` in a Flare Project Import File. `IncludePattern="*.htm;*.html;*.fltoc"` corresponds to those file types (HTM, HTML, and FLTOC) listed in **Include Files** in the Project Import Editor. Those file types will be included when the Flare Project Import occurs.

MadCap:breadcrumbsProxy

When a `MadCap:breadcrumbsProxy` is inserted into a topic, snippet, or master page, Flare creates breadcrumbs at that location in the generated content.

MadCap:conceptsProxy

When a `MadCap:conceptsProxy` is inserted into a topic, Flare creates a list of concept topics in that location in the generated content.

MadCap:endnotesProxy

When a `MadCap:endnotesProxy` is inserted into a topic, Flare creates a list of endnotes at that location in the generated content.

MadCap:glossaryProxy

When a `MadCap:glossaryProxy` is inserted into a topic, Flare creates a glossary at that location in the generated content.

MadCap:indexProxy

When a `MadCap:indexProxy` is inserted into a topic, Flare generates an index at that location in the generated content. The index may be auto-generated or based on markers in the source content.

MadCap:listOfProxy

When a `MadCap:listOfProxy` is inserted into a topic, Flare creates a list of elements at that location in the generated content. You can specify the element and class name for the list, such as `<MadCap:listOfProxy style="mc-list-of-tag: h1;" />`

MadCap:miniTocProxy

When a `MadCap:miniTocProxy` is inserted into a topic, Flare creates a mini-TOC of child topics at that location in the generated content.

MadCap:relationshipsProxy

When a `MadCap:relationshipsProxy` is inserted into a topic, Flare creates a relationship table at that location in the generated content.

MadCap:tocProxy

When a `MadCap:tocProxy` is inserted into a topic, Flare generates a table of contents at that location in the generated content.

MadCap:topicToolbarProxy

When a `MadCap:topicToolbarProxy` is inserted into a topic or master page, Flare creates a navigation toolbar at that location in the generated web-based content.

MadCapSynonyms

The root element of a MadCap Synonyms (`*.mcsyns`) file. The `<MadCapSynonyms>` element contains one or both of a `<Directional>` element and a `<Groups>` element. The `<Directional>` element contains one or more `<DirectionalSynonym>` elements with the attributes `From`, `To`, and `Stem`. `Stem` is either `true` or `false`.

The `<Groups>` element contains one or more `<SynonymGroup>` elements with a `Stem` attribute. A `<SynonymGroup>` element contains one or more `<Word>` elements.

n attribute

In Flare search in WebHelp and HTML5 output search chunk files, the n attribute of the `<stem>` element contains *n-grams*, which are sequences of characters used in a search query.

r attribute

In Flare search in WebHelp and HTML5 output search chunk files, the `<ent>` element's r attribute refers to the character position of the result in the topic.

t attribute

In Flare search in WebHelp and HTML5 output search chunk files, the `<ent>` element's t attribute refers to a topic listed in the main search file.

TocEntry

The `<TocEntry>` element represents a Table of Contents entry in a Flare TOC (`*fltoc`) file. `<TocEntry>` elements include, at least, a `Title` attribute. When there is only a `<Title>`, there is no link to a topic or other artifact. When there is a `Link` attribute, the link can point to a relative path for a Topic (`*.htm`), another TOC (`*.fltoc`), or even a Target (`*.fltar`) to be built and include as a subsystem within a help system. The `Link` attribute can also point to an image or a website. There is also an `AbsoluteLink` attribute for absolute paths.

A `<TocEntry>` can contain other `<TocEntry>` elements. This lets you create a hierarchical TOC structure.

Variable element

The `<Variable>` element represents a variable in a Flare Variable Set (`*.flvar`) file. The `<Variable>` element has a `Name` attribute and an optional `Comment` attribute. The value of the `<Variable>` is the text of the element.

w attribute

In Flare search in WebHelp and HTML5 output search chunk files, the `<ent>` element's w attribute refers to the weight or relative importance of the entry.

APPENDIX C
Glossary

.NET Framework

Pronounced "dot net framework," Microsoft's .NET Framework provides a standard library called Framework Class Library (FCL) and a virtual machine called Common Language Runtime (CLR). These are analogous to Java's Java Class Library and Java Virtual Machine. Programming languages related to .NET Framework include Visual Basic .NET and C#. MadCap Flare is a .NET application. Parts of .NET Framework are open-source software.[1]

alias and header files

These files (`*.flali`, `*.h`) enable context sensitivity in help systems. Application developers use header files to create help links in an application. Header files can be exported to a variety of formats depending on the application language. For example, a C application would used a `*.h` file while a Java application would use a `*.properties` file. Alias files are used to build header files. They are XML files and are also transformed to JavaScript in online output.

assembly

An assembly is a .NET facility that combines one or more *DLLs* in a single file, usually with a `.dll` or `.exe` suffix.

authored content

Authored content refers to content created by a human being, not by an automated process.

auto-index set

Auto-index phase sets specify phrases to include in generated indexes. However, rather than point to topics which contain index markers, index entries that result from auto-index phrase sets point to any topic that contains the phrases.

Auto-Sync

When Auto-Sync is set, Flare runs an import before running a build. Auto-Sync is the same as selecting **Auto-reimport before Generate Output** in the Project Import Editor. To set Auto-Sync, select the **AutoSync** option on the Import and Target and de-select **General → Auto-Sync → Disable auto-sync of all import files**. When Flare builds the Target, it will import files according to the rules in the Import File. Auto-Sync is also know as *Easy Sync*.

---

[1] http://blogs.msdn.com/b/dotnet/archive/2014/11/12/net-core-is-open-source.aspx

browse sequence

Browse sequences (*.flbrs) enable you to specify a reading order for topics and provide navigation that allows readers to follow that order.

C#

C# is part of Microsoft's Common Language Infrastructure (CLI). Programs written with C# can be compiled to .NET executables (*.exe) or Dynamic-link libraries (*.dll) to be used with the .NET Framework. C# is commonly written with some version of Visual Studio.[2] The syntax of C# is similar to Java.

condition tag

Condition tags let you define subsets of content for particular deliverables. For example, if you need to document multiple products that share content from a single source, you can use condition tags to include or exclude parts of your content in your deliverable to match the needs of each product or output format. Condition tag sets are maintained in *.flcts XML files.

context sensitive help

A form of help system that provides mechanisms to link relevant help topics to appropriate places in a user interface. For example, a search screen could contain a link to a help topic that describes how to use that screen. Flare provides mechanisms to create links to context-sensitive help, including URL-based and JavaScript-based methods. Topics can be aliased to provide a mechanism for selecting which topic is shown for a particular help call. This mechanism is managed with alias and header files edited through Flare's Alias Editor.

CSS

Cascading Stylesheets (*.css) let you specify output formatting and define which style classes apply to elements when editing a topic in the semi-WYSIWYG editor. CSS can be edited as text or with a Flare's CSS editor.

DLL

Dynamic-link library. A DLL[3] (*.dll) contains code and resources used by .NET applications. Dynamic-link libraries are also called *libraries* or *assemblies*. Typically, the code in a DLL is compiled from Visual Basic .NET or C# classes. Dynamic-link libraries are similar to .NET executables (*.exe), but they cannot be executed as standalone applications.

---

[2] http://msdn.microsoft.com/en-us/vstudio/hh341490.aspx
[3] http://support.microsoft.com/kb/815065

DOM

Document Object Model.[4] DOM is a W3C interface that helps to describe HTML, XHTML, and XML documents in terms of objects. When markup is deserialized into a DOM, it can be manipulated effectively with object-oriented languages such as Java or prototype-based languages such as JavaScript. Once manipulated, a DOM can be serialized into a document such as an HTML web page.

Easy Sync

See Auto-Sync.

EXE

A .NET executable file (`*.exe`) contains code and resources to be executed as a .NET application. Typically, the code in an EXE file is compiled from Visual Basic .NET or C# classes. A .NET file is not to be confused with the `net.exe` utility.

file tag

File tags (`*.flfts`) help to organize content. For example, file tags can be used to designate ownership of files by author.

Flare target

Target (`*.fltar`). Target files define the export of content to an output type such as PDF, HTML5 help, or DITA. A target file refers to a TOC or Browse Sequence that defines the content to be included in the output and its structure.

generated content

Generated content refers to content created by an automated process. This may mean content in the form of a Flare output, such as HTML5, or to Flare topics and other Flare items generated by the programmatic techniques described in this book.

glossary

A Flare glossary file (`*.flglo`) is an XML file that contains terms and definitions that can be rendered either in a central location or as pop ups in individual topics.

images and multimedia

Flare topics can contain screenshots and other images, as well as embedded videos.

---

[4] http://www.w3.org/DOM/

importing

Flare import files (*.flimp) control the import of content from Word, FrameMaker, DITA, and other Flare content. CHM content can be imported using a separate CHM import process, and external resources can be connected to Flare projects with yet another process.

index

Indexes let you identify keywords that users can use to find relevant content in a help file or other deliverable. Index entries can be specified with index markers, index link sets (*.flixl), and auto-index phrase sets (*.flaix).

index link set

Index link sets (*.flixl) specify see, see also, and sort as links in generated indexes.

Java

Java[5] is a programming language and platform created at SUN Microsystems. SUN was acquired by Oracle, which now distributes Java. Java code is compiled into Java classes and is often packaged into Java Archive (JAR) files. Java bytecode contained in Java classes can be executed by a Java Virtual Machine (JVM).

JavaScript

JavaScript[6] is a popular language for creating dynamic behavior on web pages. Web browsers include JavaScript engines that execute JavaScript code. One common use for JavaScript is *DOM* manipulation. There are also server-side JavaScript engines.

JSON

JavaScript Object Notation.[7] JSON is a somewhat human-readable format for data-interchange. It is nearly a subset of JavaScript, but it is promoted as a language-independent method for representing objects and arrays. JSON is a popular alternative to XML.

LINQ to XML

LINQ (Language-Integrated Query) to XML[8] is a feature of Visual Basic .NET and C# that lets you create and work with XML as objects. Using Visual Basic .NET, XML can be coded as XML literals[9] which are converted to objects at run-time.

---

[5] https://www.java.com/en/
[6] https://developer.mozilla.org/en-US/docs/Web/JavaScript
[7] http://www.json.org/
[8] http://msdn.microsoft.com/en-us/library/bb397926.aspx
[9] http://msdn.microsoft.com/en-us/library/bb397926.aspx

MadCap MadPak

Flare is the centerpiece of the MadPak suite of tools. But there are also tools, each XML-based, for managing and layering images (Capture), software simulation (Mimic), and translation (Lingo). Analyzer is a scaled up version of Analyzer features in Flare that uses scanning and a database to present information about Flare projects.

Many of the XML manipulation techniques discussed in this book can also be applied to the XML artifacts used by other products in the MadPak suite.

master pages

Master pages (`*.flmsp`) are used in web outputs to wrap topics and contain content common to every topic in an output. Master pages are XML files with XHTML-like content.

multichannel output

Output content in multiple forms, such as PDF and Flare HTML5 help, that is generated from a single source.

n-gram

In Flare search in WebHelp and HTML5 outputs, an n-gram is sequence of characters that may be contained in a search query. N-grams are stored in the n attribute of the `<stem>` elements in the main search data files and the search chunk files. The `<stem>` element's n attribute links the main search file and the chunk files.

page layout

Page layouts (`*.flpgl`) are used in print outputs. Page layouts can specify distinct layouts for several types of pages including First, Left, Right, and Empty.

regular expression

A regular expression is a string of characters used to search for patterns. For example, in a search, the regular expression string "cat" would match the string "cat". What makes regular expressions more interesting than just a direct match are wildcard characters. These come in many varieties, but commonly a "?" will match any single character and a "*" will match 0 or more characters. So, "c?t" would match "cat" but not "bat" and "c*t" would match "ct", "cat", and "cart" along with many other possibilities.

relationship table

Relationship tables (`*.flrtb`) in Flare are similar to DITA relationship tables. They establish connections among related topics. Relationship tables are typically displayed as a list of links at the end of a topic.

report

Flare reports (*.flrep) provide information about content in a project gathered by Analyzer. For example, there is a report for broken links. Analyzer scans projects and stores information about those projects in a database. Analyzer is also a separate application that does the same thing but with more options.

search chunk

The data files for Flare's WebHelp and HTML5 search functionality include a main search file that references one or more search chunk files. These XML and JavaScript files (the JavaScript file contains the same information as the XML file, but wraps that information in a JavaScript function) contain information about each stem, which are represented by a sequence of characters called an *n-gram*. The stem entry in the main file indicates which chunk file contains more information. The stem entry in the chunk file indicates how to rank results, contains a key to the topic result, and identifies where the stem occurs in the topic.

search file (main search data file)

This file is the point-of-entry for search data in WebHelp or HTML5 outputs. There is an XML version and a JavaScript version. The JavaScript version contains an escaped copy of the XML version as a string value for a variable. The main search data file references one or more search chunks that contain additional information about what to return for searches containing various sequences of characters.

search filter set

Search filter sets (*.flsfs) are XML files that define search filters in web-based output. A search filter appears as a drop down connected to the search field.

single source

An individual source file, set of source files, or other data source that is used to produce more than one output type. When content is dynamically rendered, each output type may be rendered differently. Examples of source files used by Flare include topics and TOCs.

skin

Skins (*.flskn) define the structure and look of a web-based help output. For example, a skin might display TOCs in a navigation pane on the left side and topics in a content pane. A skin also defines where other elements, such as indexes and glossaries, are displayed.

snippet

Snippets (*.flsnp) are similar in structure to topics and master pages. They are XML files with XHTML-like content. Snippets are discrete chunks of formatted content that can be included in topics and master pages.

solution

Solutions are a container used to manage Visual Studio projects. They include all of the files and metadata for one or more projects.

table style

Flare provides a CSS editor that creates a separate CSS (*.css) file just for tables. The Flare CSS editor provides a set of table-related features. However, you can also define styles for tables in the main CSS file.

TOC

TOCs (tables of contents) are XML files (*.fltoc) that contain label and path information for topics or several other types of TOC entries. Topics and other entries can be added, moved, and deleted with the TOC editor in Flare.

topic

Topics (*.htm) are the basic unit of content in Flare. Topics use XHTML markup with MadCap-specific extensions. Topics can be edited as text or using Flare's semi-WYSIWYG topic editor. Topics can be organized hierarchically into TOCs.

Topic Editor

Flare's Topic Editor provides an XML Editor and a Text Editor. The Text Editor displays markup with syntax coloring. The XML Editor displays the text of the markup and blocks on the left and top to represent tags. Both views can be shown at the same time with synchronized updates.

variable

Flare lets you create variable definitions for use in your content. Variable definitions are maintained in *.flvar XML files. You can create a variable set that contains multiple variables. For example, you could have one set of variables with US spellings and another with UK spellings to deliver the same project to different locales. Flare variables are globally accessible in a project.

Visual Basic .NET

Visual Basic .NET[10] is part of Microsoft's Common Language Infrastructure (CLI). Programs written with Visual Basic .NET can be compiled to .NET executables (`*.exe`) or Dynamic-link libraries (`*.dll`) to be used with the .NET Framework. Visual Basic .NET is commonly written with some version of Visual Studio. The syntax of Visual Basic .NET is similar to older versions of Visual Basic.

Visual Studio

Visual Studio is an integrated developer environment created by Microsoft. There are multiple versions available. A Flare plugin would normally be programmed in Visual Studio.

XML

Extensible Markup Language.[11] XML is intended to be human and machine-readable. XML is text annotated by tags. XML can be validated against XML schemas to ensure compliance to a set of rules defined by the schema. XML is standardized by W3C. Many MadCap Flare files are XML.

XSLT

Extensible Stylesheet Language Transformations.[12] XSLT is a language that lets you transform a source XML document into a target XML document or into other forms, including text.

---

[10] http://msdn.microsoft.com/en-us/library/2x7h1hfk.aspx
[11] http://www.w3.org/XML/
[12] http://www.w3.org/Style/XSL/

About the Authors

Thomas Tregner

Thomas Tregner was raised in Villa Park, Illinois. A secondhand Commodore 64 and some encouragement from his father gave him his first exposure to programming. After graduating from high school, Thomas worked at various jobs and took college courses before enlisting in the US Navy. He graduated from the Navy Nuclear Power Program and served on USS Enterprise.

After his time in the Navy, Thomas worked as a technical writer while finishing a Bachelor of Science in Information Systems. He then transitioned to programming. He is currently a software engineer at a publicly held software company in Charleston. Thomas also holds a Master of Science in Computer Information Systems from Nova Southeastern University. He has a blog about this book's subject matter at www.tregner.com.

Thomas and his wife, Laura, live Charleston, South Carolina, with their three children.

David Owens

David Owens is Group Manager of User Education at Blackbaud, the leading provider of software and services for nonprofit organizations. As Group Manager, David leads several distributed feature documentation teams and an SDK documentation and training team. The group includes technical writers, information architects, and software developers who create content that blends traditional and non-traditional user assistance formats. His teams have been using MadCap Flare as their primary authoring tool since 2011.

As a technical writer and information developer, David has created feature and API user assistance content in a wide variety of formats for software products ranging from an enterprise fundraising CRM application and a business intelligence multidimensional data analysis system, to shrink-wrapped mass market software, as well as commercial hardware. David has published articles on topics such as single-sourcing and how to become a "technical" technical writer. He also contributed a section to *Troubleshooting Microsoft Technologies: The Ultimate Administrator's Repair Manual* published by Addison-Wesley.

https://www.linkedin.com/pub/david-owens/8/735/663

Index

Symbols
.NET, 37
    namespaces in, 104

A
actions, connecting Toolbar functions to, 80
Adobe FrameMaker (*see* FrameMaker)
alias files, 93–95
Analyzer, 11
API
    Adobe Acrobat, 83
    DOM manipulation with the plugin, 121
    DotNet Help, 97–98
    manipulating documents with the plugin, 118
    plugin, 39, 99–124
    plugin code example, 117
    reference, building, 47
applications, connecting Flare help to, 91–92
artifacts, 10
assembly, 99
authored content
    combining generated content with, 131–135
authoring
    Flare editors for, 19–21
    topic-based, 13–15
authoring tool, Flare as, 16
Auto Index Sets, 66–69
Auto-suggest, 11
Auto-Sync, 16–18, 85–86
    procedure to use, 17
automation, document, 83

B
B3.PluginAPIKit.dll, 99
batch file generator, 86–89, 139
batch files, 2, 83
    create search tests using, 77
    templates for, 84
Batch Target Editor, 83
bookmarks, 26
breadcrumbs
    generated by proxies, 28
    specifying, 49
browse sequence, 45
builds
    automating, 83
    parallel, 86
button
    connecting Toolbar functions to a, 80
    creating in toolbar and ribbon UI, 111–115
    editor information plugin, 115
    open help topic using a, 93
    sort TOC with JavaScript, 59–60

C
C#
    class file example, 101
    code for creating UI in version 10, 112
    code for creating UI in version 9, 114
    connecting to Flare help, 96
    console example, 38
    IPlugin interface example, 106
    tool to create multiple text targets, 77
    Visual Studio Express and, 100

CHM, 97
   import wizard, 16
class attribute
   defining type with, 14
   JavaScript to highlight text based on the, 40–42
   styling using the, 21–23
   XSL code to match on the, 36–37
CLR (common language runtime), 99
code samples
   cutting and pasting, 126
   formatting, 125–126
color, syntax highlighting using, 125–126
common language runtime (CLR), 99
conditional content, 15, 27
content reuse, 13
copy and paste, 126
csh.js, 93
CSS, 21–23
   JavaScript fragment using, 41
   setting button look and feel using, 80

D

Darwin Information Typing Architecture (*see* DITA)
DITA, 4
   compared with Flare, 15
   output, 18, 31
   topic import, 31
   topic-based authoring and, 13–14
DLL (dynamic-link library), 99
document automation, 83
documents
   manipulating with the plugin API, 118
DOM (Document Object Model), 39
   Flare support for manipulating the, 107, 116
   using a plugin to manipulate the, 121
DotNet Help, 96–97
drag and drop
   creating a TOC with, 47
   topics, 21
dynamic-link library (*see* DLL)

E

Easy Sync (*see* Auto-Sync)
Eclipse Help, 98
editor plugin, 115–118
element tags, MadCap proprietary, 26
embedded help, 98
exports
   Flare, 17
   table of contents, 48
external help, 46
External Resources feature, 32

F

file types
   .flali, 93–95
   .flbrs, 45
   .flglo, 70
   .flimp, 16
   .flimpdita, 31
   .flimpfl, 31
   .flixl, 66
   .flmsp, 23
   .flprj, 7
   .flsfs, 75
   .fltar, 10
   .fltoc, 19, 43
filters
   conditional text, 27
   search, 75
find and replace, 33
Flare.exe, 99
folders, Flare, 8–11
fonts
   code, 125
   plugin to collect information about, 115
FrameMaker
   Auto-Sync and, 18
   import mapping for, 46
   importing, 31
functions, anonymous, 79

G

generated content
    strategies for managing, 131–135
glossaries, Flare, 70–72
glossary proxies, 70
granularity, 13

H

header files, 93
    aliases defined in, 94–95
help viewer, redistributable, 97
helpers, JavaScript, 93–95
highlighting, syntax, 125–126
HTML Help, 46, 97
HTML5, 91, 97
    bookmark target for, 46
    JavaScript helpers, 93
    navigation skins in, 28
    syntax highlighting, 126
    tables of contents with, 48
    topic output, 33

I

IDE (integrated development environment)
    formatting code with an, 125
iframes
    HTML5, 91
    manipulating, 33
import, 31–32
    Auto-Sync, 16–17
    automating, 85–86
    table of contents, 46
Import Editor
    defining import mappings with, 46
index proxies, 66
indexes, 65–69
    Auto Index Sets, 66
installation, default location for Flare, 24
instance properties, 106–107
IntelliSense, 104

J

Java
    accessing help from, 95–96
    creating Flare topics and TOCs with, 60–63
JavaScript, 79–81
    accessing search data using, 73
    create a custom filter with, 33
    create button to sort TOC with, 59–60
    helper functions in, 93–95
    in topics, master pages, and snippets, 81
    sort HTML5 glossary using, 71–72
    table of contents generated with, 49
    underline content based on class attribute, 40–42

L

LINQ to XML, 37–39
logging, 83

M

madbuild.exe, 83
MadCap, proprietary element tags, 26
MadCap.OpenHelp, 93
master pages, 23
    JavaScript in, 81
    proxies and, 28–30
    XML tags in, 26
menus
    assemblies required to implement, 102–103
    creating a plugin with, 111–115
    types of, 1
Microsoft Word
    Auto-Sync and, 18
    compared with Flare, 18
    import mapping for, 46
Mimic, 46
mini-TOCs
    proxy for, 28–30
    specifying, 49
multichannel output, 15, 20

N

namespace
  MadCap, 26
namespaces
  .NET, 104
navigation
  browse sequences for, 45
  print, 52
  using proxies for, 28
  web-based output, 48–52
nesting
  tables of contents, 44
New Project Wizard, 8

O

output
  DITA, 31
  types of, 30–31
  web-based, 48

P

parallel builds, 86
Pascal, accessing help from, 96
PHP, accessing help from, 96
plugins, 37, 99–124
  (*see also* API)
  building and deploying, 109–111
  creating menus in, 111–115
  editor information, 115
  Flare support for, 107
  using Visual Studio to create, 100–115
prettify, syntax highlighting with, 126
preview, XML Editor, 20
print output, 31
  table of contents, 52
Project Organizer, 10–11
projects, 7–12
  building, 45
  managing content in separate, 132
  New Project Wizard, 8

properties
  setting button, 80
  viewing Flare, 21–23
properties files, 96
proxies, 28–30
  appearance of in print, 52
  appearance of in web output, 48
  glossary, 70
  index, 66
  search, 72
proxy-based navigation, 49

R

redistributable help viewer, 97
regular expressions, 34
relationship tables, 15
RequireJS, 49
reuse, 13
ribbon UI
  building a plugin for the, 111–115
root folder, 8

S

scheduling, batch, 83
schema
  Flare topic, 24
  location of Flare, 43
  MadCap Flare, 10
  MadCap.xsd, 26
scripts (*see* JavaScript)
search, 33
  customized, 72–75
  data file structure for, 73
  proxy, 72
  URL-based, 92
search filters, 75
single source, 15–16
Skin Editor, JavaScript and the, 79
skins
  customizing search with, 72
  navigation, 28

snippets, 26–27
    generating content as, 134
    JavaScript in, 81
specialization, DITA, 13
Spellcheck, 11
split view, Flare editor, 20
Styles window, 21
stylesheets, for multichannel outputs, 15
    (*see also* CSS)
syntax coloring, 125–126

T

table of contents, 19
    example of, 2
    exporting, 48
    file type, 19
    manipulating, 47–48
    nested, 44
    print output, 52
    Project Organizer, 10
    proxy for, 28–30
    structure of, 43–45
    TocEntry element, 44
    web-based, 48–52
targets
    batch, 83
    folders for, 11
Team Foundation Server, 83
    integrating Visual Studio with, 84–85
templates
    batch file, 84
    topic, 32
Text Editor, 20
text, conditional, 15
TOC (*see* table of contents)
TOC Label, 21
TocEntry, 43
Toolbar JavaScript, 79–81
Topic Editor, 11
    interaction between plugin API and, 116
topic templates, 32

topic-based authoring, 4, 13–15
topics
    authoring Flare, 14–15, 19
    code samples to modify, 33
    combining generated and authored, 134
    comparison, 16
    contents of, 19, 26
    defined, 13
    defining new types of, 22
    DITA, 13–15
    JavaScript in, 81
    modifying, 33
    overview, 19–21
    predefined, 22
    structure of, 24–25
    styling, 21–23
    types of, 13
    XML markup for Flare, 24
transclusion, 132

U

URL syntax, 91–92
URL-based search, 92
user interface
    customizing, 99–124
    Flare, 1
    plugin, 104
    working outside the, 2

V

variables, 28
video screencasts, 46
Visual Basic, 38
    accessing help from, 96
    TOC sorter using, 56
Visual Studio, 38
    creating a Flare plugin with, 100–115
    syntax coloring in, 125
    Team Foundation Server and, 84
Visual Studio Express, 100

web output, 31
    table of contents, 48
WebHelp, 97
whitespace, syntax highlighting using, 126
wizard
    import, 16
    New Project, 8
    table of contents, 19
Word (*see* Microsoft Word)

XHTML
    element tags, 26
    Flare topics and, 25–26
XML declaration, 24
XML Editor, 19
XML schema (*see* schema)
XSLT, 34–37
    TOC example using, 53

About XML Press

XML Press (http://xmlpress.net) was founded in 2008 to publish content that helps technical communicators be more effective. Our publications support managers, social media practitioners, technical communicators, content strategists, and the engineers who support their efforts.

Our publications are available through most retailers, and discounted pricing is available for volume purchases for business, educational, or promotional use. For more information, send email to orders@xmlpress.net or call us at (970) 231-3624.

CPSIA information can be obtained at www.ICGtesting.com
Printed in the USA
LVOW05s1557020715

444727LV00004B/6/P